FLOODS AND DRAINAGE

LIBRARY
AVERY HILL COLLEGE
Bexley Road, Eltham,
SE9 2PQ.

FLOODS AND DRAINAGE

British policies for hazard reduction, agricultural improvement and wetland conservation

E. C. Penning-Rowsell
D. J. Parker, D. M. Harding

London
ALLEN & UNWIN
Boston Sydney

**Allen & Unwin (Publishers) Ltd,
40 Museum Street, London WC1A 1LU, UK**

Allen & Unwin (Publishers) Ltd,
Park Lane, Hemel Hempstead, Herts HP2 4TE, UK

Allen & Unwin, Inc.,
8 Winchester Place, Winchester, Mass. 01890, USA

Allen & Unwin (Australia) Ltd,
8 Napier Street, North Sydney, NSW 2060, Australia

First published in 1986

ISSN 0261-0507

British Library Cataloguing in Publication Data
Penning-Rowsell, Edmund C.
Floods and drainage: British policies for hazard reduction, agricultural improvement and wetland conservation. – (The Risks & hazards series, ISSN 0261–0507; v. 2)
1. Drainage – Great Britain
I. Title II. Parker, Dennis J.
III. Harding, Donald M. IV. Series
627'.54'0941 TC978.G7
ISBN 0–04–627001–9

Library of Congress Cataloging in Publication Data
Penning-Rowsell, Edmund.
Floods and drainage.
(Risks & hazards series, ISSN 0261–0507; 2)
Bibliography: p.
Includes index.
1. Flood control – Government policy – Great Britain.
2. Drainage – Government policy – Great Britain.
3. Wetland conservation – Government policy – Great Britain. 4. Agriculture and state – Great Britain.
I. Parker, Dennis J. II. Harding, Donald M.
III. Title. IV. Series.
HD1676.G7P46 1985 363.3'4936'0941 85–13383
ISBN 0–04–627001–9 (alk. paper)

Set in 10 on 11 point Times by Columns, Reading
and printed in Great Britain by Mackays of Chatham

Doctor Foster
Went to Gloster
In a shower of rain.
He stepped in a puddle
Right up to his middle
And never went there again

Anon.

This book is dedicated to Dr Foster who, by retreating in adversity, happily left our research field wide open.

Foreword

The *Risks and Hazards Series* is designed to make available to a wider readership the research results and intellectual developments arising from current concerns with hazards in the environment, and the associated risks to human health and wellbeing. Scientific as well as public concern about hazards has reached a new peak but the changes occurring are more than a matter of degree.

There is in all Western industrial societies a new element of public anxiety about hazards, which threatens to erode the confidence and trust that people place in the most powerful institutions. Because of this anxiety and the common policy and management questions that arise, hazards are being studied for the first time as a set of related phenomena. This generic approach is leading to an erosion of the commonsense distinction between natural and man-made hazards. Across a wide spectrum of risks, many people are no longer content to accept the inevitability of adverse effects as being natural or to be expected. Each event requires an explanation, and often the search for explanation is linked to a search for the locus of responsibility – someone to blame. These new circumstances have been in part generated by intellectual enquiry, but in their turn they are now posing new questions for research and giving rise to a demand for more effective policies.

Two decades ago systematic enquiry into the causes and consequences of natural hazards was conducted in disciplines and professions that worked to a considerable degree in isolation from each other and from the public. Beginning in the 1960s a much broader interdisciplinary approach to hazards developed, strongly influenced by geographers (natural hazards research) and by sociologists (disaster research). This research has shown that the consequences of floods and droughts can no longer be explained away as 'Acts of God' but are often the result of the misapplication of technology. More recently the realisation is emerging that the technical shortcomings of particular human agencies are not the whole story either. To understand fully the problem of environmental hazards we must delve deeper into man's relationship with nature and the nature of human society.

The time seems opportune therefore to launch a new series that will provide more effectively for the substantial development and presentation of the empirical and conceptual advances that may be expected. In keeping with the nature of the subject matter, the series is planned to be international and interdisciplinary in scope and will deal with both natural and man-made hazards. As a vehicle for major contributions from scholars and practitioners the series is intended to meet needs in three directions: (a) professional managers, (b) research workers, and (c) students and teachers.

The social and scientific problems associated with environmental hazards now command the attention of a large group of management professionals in public and private organisations at local, national and international levels. The series is expected to help identify new directions in research and policy as the understanding of the causes and consequences of hazards deepens.

The interactions between hazard theory, research and management practice are now being incorporated into the content of university-level courses in the natural and social sciences as well as a number of interdisciplinary fields such as environmental studies and planning.

The common theme of the series is the concept of hazards and their management, and the orientation will be largely towards practice and policy. More theoretical works and 'critical science' approaches are not excluded, and the series will be especially pleased to consider manuscripts that arise from interdisciplinary work; that consider more than one hazard; or that study hazards in a multinational or cross-cultural context; and present a broad human–ecological approach to environmental hazards.

What the authors of this second volume in the *Risks and Hazards Series* refer to as 'conventional hazard–response theory' has tended to give weight to individual behaviour in the face of environmental hazards. Continuing the critical viewpoint adopted by Kenneth Hewitt and his co-authors in *Interpretations of calamity*, the three authors of *Floods and drainage* show both the conceptual and practical importance of institutional and political factors in hazard management.

They achieve this objective by an in-depth analysis of flood and related drainage problems in the United Kingdom. The very fact that flood problems cannot be adequately studied in Britain without close consideration of the politics and institutional arrangements for land drainage goes a long way to demonstrate their argument. In other parts of the world, notably in much of the United States, the word 'drainage' is more commonly coupled with irrigation than with floods. Edmund C. Penning-Rowsell, Dennis J. Parker and Donald M. Harding thus in one step broaden our conception of flood problems. In keeping with the tradition and intent of much conventional hazard research, they show that broad multidisciplinary approaches are necessary to a satisfactory formulation of environmental hazard problems in contemporary society.

The authors are not content to have made an important conceptual contribution. They provide guidance by example. By linking the methods of institutional, hydrological and economic analysis, they show in practical terms the consequences of their conceptual innovation. Two decades ago an analysis that effectively linked institutional, hydrological and economic analyses would automatically be at the forefront of the field. This is no longer the case. Now it is important to recognise environmental and social impacts as well. The inclusion of these in terms of public involvement demonstrates how rapidly the thinking and practice of land drainage has progressed in two decades.

By bringing these factors together in a common framework Penning-Rowsell, Parker and Harding provide an up-to-date perspective on the practice of flood alleviation and land drainage.

Although of prime interest to students and professionals in the United Kingdom, what they have to say will influence thinking much more widely.

IAN BURTON

Preface

This book has developed from 15 years' research into flood hazards at Middlesex Polytechnic and a period before that at University College of Wales, Swansea. During this time the Middlesex Polytechnic Flood Hazard Research Centre has developed a reputation for researching and applying methods of benefit–cost analysis in the fields of flood alleviation and agricultural drainage. This work has been undertaken principally by geographers committed to applying their skills to practical environmental and social problems. The results of the research have been disseminated through a large number of papers, books and consultancy reports as well as through professional training courses and many informal contacts.

Over the past 15 years the research has been supported by funds from Middlesex Polytechnic, the Natural Environment Research Council, the Ministry of Agriculture, Fisheries and Food, the Central Water Planning Unit as well as several Water Authorities and local authorities. During this period we have developed strong links with North American and Australian hazard researchers and practitioners, but perhaps our most regular links are with practising land drainage engineers in Britain. Our experience in applying our research and methods has inevitably led us beyond the area of flood alleviation and drainage economics and into the fields of engineering, hydrology, environmental management and public consultation. This diversification is reflected in the objectives and content of this volume.

Our first objective in this book is to advance hazard–response theory as developed mainly by North American geographers of the Chicago 'School' dominated by White, Burton and Kates. Much of this theory has emphasised personal response to hazardous environments and notably to flood hazards. We have become increasingly dissatisfied with the inapplicability of much of this emphasis in conventional hazard–response theory to the floods and drainage field as we know it in Britain. Individual responses here are vital in making flood warnings effective or in generating sufficient local willingness to pay for flood alleviation. However, the over-concentration on individual response alone seriously detracts from the institutional and political forces which, in our view, are the key to analysing hazard responses or policies and their shortcomings.

In Chapter 1 we therefore outline our conceptual view of the institutional and political influences which, in our experience, are of central importance in analysing flood hazard mitigation and drainage improvement policy. At the end of Chapter 1 we also, briefly, bring the reader up to date with certain important recent developments (A developing scene). Chapter 2 continues with an analysis of the British institutional arrangements in the spheres of flood alleviation and agricultural drainage improvement, which have a common institutional framework as well as significant hydrological interlinkages. The analysis in the following chapters is based upon our conceptualisation of the factors and inputs affecting policy decisions. These chapters therefore represent an institutional and policy analysis of the influences on these decisions; they are therefore relevant to both the student of environmental studies and the practising drainage engineer. Our emphasis on institutional analysis does not replace the individual mode of hazard–response theory in any simple way but rather the two are complementary. Different theories have different value and applicability in analysing particular problems and policies in contrasting circumstances.

Our second objective is to present a volume of broad and comprehensive practical relevance to professionals within the field of flood alleviation and land drainage, particularly in Britain, rather than to report the detail of research results *per se*. The effective management of flood and drainage problems needs this type of comprehensive analysis now more than ever before. Experience shows that single-discipline approaches, whether engineering, economic, hydrological, social or environmental, are destined only to be partially successful and indeed sometimes to be counter-productive. Examples might include the single-minded pursuit of structural flood mitigation engineering solutions in North America and the neglect of environmental values in many drainage proposals in Britain (Ch. 5). Progress, therefore, appears to lie in the direction of comprehensive approaches in which all aspects are brought together into a common framework of analysis: this requires both new institutional arrangements and new modes of analysis.

A more comprehensive approach places new and greater demands upon the land drainage engineer who must master a wide range of skills. We therefore bring together in this volume several interrelated topics including the institutional context (Ch. 2) and hydrological or hydrographic data inputs to project appraisal (Ch. 3), as well as the economic (Ch. 4), the environmental (Ch. 5) and the public involvement (Ch. 6) aspects of flood and drainage problems. The aim is to give the practising professional – as well as the student – an up-to-date perspective over the wide range of knowledge which makes up the interdisciplinary field of modern flood alleviation and land drainage planning. Where possible we have attempted to identify key lessons and particular areas for further research.

Despite the long period of research leading towards this book our understanding of the forces moulding policy decisions is nevertheless in its infancy. Thus our predictions for the course of events and policy evolution will remain imprecise if not sometimes clearly wayward. The continuing research at Middlesex Polytechnic is undoubtedly helping at the margin. This research currently focuses on technical aspects of benefit–cost evaluation – specifically on indirect and 'intangible' benefits – and serves to show how different groups gain and lose by investment decisions in this sphere. However, only with further detailed analysis of case examples and the history of policy evolution will our emphasis on the structural forces in society influencing policy evolution take both clearer form and more detailed content as our knowledge deepens. There remains much to be done.

EDMUND PENNING-ROWSELL
DENNIS PARKER
DON HARDING
July 1984

Acknowledgements

We would like to acknowledge the contribution of many individuals to the production of this volume. First and foremost we are indebted to our colleagues John Chatterton, Roger Witts, Colin Green and Paul Thompson. We must also acknowledge the contributions from our other colleagues in the School of Geography and Planning at Middlesex Polytechnic to the intellectual environment in which this research and writing has taken place. In particular we are indebted to Madeleine Wahlberg, Dennis Hardy, Aram Eisenshitz, Paul Smith and Annabel Coker. In addition, Steve Chilton has assisted with the illustrations and Michele Smith with the manuscript. We are indebted to Mrs M. Penning-Rowsell for her sterling proof-reading work and for the creation of the index. We also acknowledge the permission for reproducing diagrams given by the Natural Environment Research Council, Laurence Gould Consultants Ltd, the Water Authorities Association, the *Journal of Environmental Management*, the Local Government Operational Research Unit, R. W. Kates, Longman Group Ltd, John Wiley and Sons Ltd, and the International Disasters Institute. Laurie Greenfield and Carol Murdoch have helped with bibliographic advice and Annabel Coker gave valuable assistance with material in Chapter 5. Malcolm Hewson, Adrian MacDonald, Ian Kelso, Alun Hughes and Roy Ward gave useful advice on some material in Chapter 3 and Roger Buckingham and Tony Hughes provided data for Chapter 2. Peter Thorpe gave assistance with material for Chapter 6. However, the authors alone are responsible for the text as it stands. Timothy O'Riordan gave unstinting friendship and advice over the whole field over a number of years. Finally we would like to acknowledge the considerable help given at crucial stages by Ian Burton and also the patience of Roger Jones.

Figures 3.4, 4.5, 5.5 and 6.2 are reproduced from the Ordnance Survey maps with the permission of the Controller, Her Majesty's Stationery Office: Crown copyright reserved.

Contents

List of figures

List of tables

1 *Flood alleviation and agricultural land drainage*

Introduction

Urban flood alleviation and agricultural land drainage are inseparable in Britain. In many parts of the country the soil–water regime inhibits agricultural use. Increasing agricultural production here depends, therefore, upon the provision of adequate field drainage. This in turn depends on having adequate discharge capacity in the network of interconnecting drainage channels and rivers downstream. These rivers and channels often pass through urban areas and can cause flooding. Towns and cities in turn may discharge their rapid storm runoff through further agricultural areas, again perhaps causing flooding or preventing field drainage.

Both urban and rural flood alleviation or drainage must therefore be viewed in concert because the one may exacerbate the other's problems. As well as being hydrologically interdependent, the two aspects of drainage are also institutionally linked in Britain. This reflects many decades of experience of these hydrological interlinkages and thus a keen appreciation of the advantages of undivided field-to-coast river catchment management by single land drainage agencies.

Our complementary analysis of policies to reduce flood hazards and agricultural drainage problems emphasises that the management of such hazards is concerned with the utilisation or exploitation of natural resources. A fundamental objective of flood hazard reduction and land drainage schemes is thus improving the use of land. The process is complex, however, in that the balance of environmental systems is being disturbed. Consequences may be inadvertent and unpredictable. Drainage schemes may enhance land values for one purpose but devalue the same land for other uses. Flood hazard management therefore involves complex trade-offs between alternative resource gains and losses.

Flood alleviation and land drainage are fundamentally about planning public expenditure to increase social welfare. Urban and agricultural drainage schemes are designed to improve the nation's welfare in much the same way as are hospitals, education and the social services. The aim is to reduce flood damage, enhance agricultural productivity and thereby to promote greater prosperity by ensuring the security of property, a healthy workforce, and efficient business, including farming, through the provision of adequate infrastructure including drainage.

This volume aims, in this context, to elaborate on the inseparability of flood hazard reduction and agricultural land drainage in Britain, with all the complications – and interest – that this linkage brings to our analysis of policies and decision making. Secondly, we aim to show that a multidiscipli-nary approach to policy analysis and problem solving is necessary for progress. For appropriate policies and good management practices those making decisions need to combine theories and information from environmental science, engineering, institutional analysis, economics, decision making and planning.

A third aim is to attempt to fit hazard reduction and land drainage into a wider analysis of the relevant political economy of Britain today. This is necessary because the policies being pursued are inseparable from this wider context and cannot properly be understood in isolation. Through pursuing these emphases we aim, fourthly, to contribute a new perspective on hazard–response theory which brings together into a common framework the analysis of two complementary hazards and emphasises the institutional and political influences upon the decision-making process.

Flooding and land drainage problems in Britain: the environmental context

Britain has a temperate cyclonic climate giving predominantly winter rainfall averaging some 1000 mm per annum (Manley 1952; Meteorological Office 1973; Rodda *et al.* 1976, Chandler & Gregory 1976). Intensive storms and floods are caused by westerly depressions with well-developed warm fronts or by summer convectional thunderstorms (Ward 1975, 1978, Smith 1972). Rapid winter snow-melt in changeable weather regimes can also produce serious flood conditions. High levels of water retention in lowland clay soils, together with the low evapotranspiration rates in a cool moist climate, result in a relatively long season of soil waterlogging. The extensive low-lying areas in eastern England with naturally high water tables also present continual drainage problems. Large and extensively paved urban areas, again particularly in England, can produce rapid runoff and sudden floods (Hollis 1975, 1979). Because catchments are relatively small, most British rivers tend to be 'flashy' in nature giving very little time for flood warnings (Lewin 1981).

Although they are diverse, most British flood and drainage problems are perhaps relatively minor by world standards; fortunately loss of life in floods is comparatively rare. Nevertheless Britain has a long tradition of both flood alleviation and agricultural land drainage (Johnson 1954, Cole 1976). Analytical techniques have developed further than in most countries, and institutional arrangements have evolved and matured over many generations. The result is that annual expenditure in this field is currently about £80 million (1981–2). The ratio of expenditure on agricultural and urban drainage improvement is approximately 40 : 60 and nearly all is for England and Wales. Scotland suffers from few major flood or drainage problems, despite a wetter climate, because a smaller population means less pressure on land and less intensive use of floodplains.

The causes of British urban flood problems are varied. The historic cores of many of the older British towns were located adjacent to river floodplains to capitalise upon defensive locations and natural lines of communication. Subsequent development during periods of urban expansion has tended to be on flood-prone land (Penning-Rowsell & Parker 1974). Such floodplain encroachment occurred principally during the 19th century, but it also occurred in the 20th century before the Town and Country Planning Act 1947 created a universal land-use planning system to control urban development (Penning-Rowsell 1981a). Although this legislation was not specifically enacted to control floodplain encroachment, the Act and subsequent government circulars (Ch. 2) have largely been successful in limiting extensive development of floodplains. The hazards from pre-existing encroachment have remained, however, in those locations where structural flood alleviation schemes have not been implemented or are of insufficient standard. Urban floodplain encroachment continues as a nagging but ill-defined problem, although it is on nothing like the scale found in the United States where general floodplain regulations or zoning systems have yet to be adopted (Parker & Penning-Rowsell 1983).

Besides major urban encroachment of floodplains resulting from the pre-1947 urban expansion, as in the lower parts of the Thames catchment west of London, numerous smaller flood problems also exist. Many arise because British urban drainage systems are commonly decades or centuries old, highly modified and inadequate for today's increased runoff from dense urban development and improved agricultural land drainage. Urban watercourses have commonly been culverted or piped underground and combined with underground storm drainage and sewerage systems which may now be inadequate to discharge peak flows. The result is disruptive flooding from unpredictable storm runoff, the reduction of which will be costly in terms of improved or replacement drainage capacity.

Large areas of eastern England are close to or below sea level and are comparable in some cases to the polderland areas of the Netherlands (see Fig. 2.3). Major drainage systems have been developed during the past 200 years. At first these were powered by wind pumps and later by steam, diesel and electric pumps in conjunction with thousands of kilometres of artificial arterial drainage channels and dykes (Darby 1940, 1983). Water tables have been lowered so that bog, mere and marsh soils have been converted into more valuable agricultural land by capitalising upon the reduction in flooding frequency. English drainage expertise – itself partly learned from Dutch engineers – originated largely in these Fens but wetlands elsewhere in the country have been drained progressively to improve agricultural productivity.

Field drainage is now considered a basic requirement for efficient farming on over half of the 11 million hectares of agricultural land in England and Wales. Cole (1976) suggests that the use of a further 3 million hectares is constrained by poor drainage and that completing the necessary drainage improvements for this land will require major river improvement works. However, the environmental impact of continual agricultural and urban drainage improvements has become a major environmental issue in Britain. Wetland habitats have become progressively scarcer and therefore increasingly valued (Penning-Rowsell 1980, 1983a).

Because so much of eastern England is low-lying, sea flooding is a major problem, especially from the Humber estuary to Kent (Steers 1953, Ward 1978). The threat is from a combination of high tides, low pressure over the North Sea and strong onshore winds which lead to tidal surges. Disastrous floods occurred in 1953 when 300 people were killed and property losses amounted to some £30 million (approximately £250 million at 1983 prices). Unfortunately eastern England is falling relative to sea levels and the 1953 event came close to being repeated in 1978, thus emphasising the need for continual re-evaluation of east coast sea defences against a growing risk. Raising these defences, however, requires massive expenditure and new technology in the form, for example, of tidal exclusion barriers such as those installed in the Thames estuary to protect London (Horner 1979). Decisions must therefore involve the most careful evaluation of the risks, costs and the adverse impacts of such ventures.

The welfare products of flood alleviation and land drainage

Flood alleviation and land drainage are fundamentally about planning public expenditure to increase social welfare by reducing the type of flooding and drainage problems described above. This raises questions concerning why public expenditure is necessary in this field – why should the state be involved? – and how should decisions be made concerning social welfare? In tackling these questions we have to disentangle the rhetoric of government and policy makers from the reality as experienced by those affected by flooding and drainage problems and by those involved in planning the drainage improvements.

THE CASES FOR STATE INTERVENTION

Much research into human response to floods has its origins in the Chicago school of geographers who primarily emphasised the role of the individual in making adaptation decisions to counter the hazards they face (Kates 1962, 1970, Burton *et al.* 1978). Nevertheless in Britain, as in most countries, the dominant flood alleviation and land drainage policies are government sponsored and supposedly oriented towards achieving community goals. Raising overall social welfare – the product of all individuals' welfare – is the objective and state intervention and community funds can therefore be justified. Given the free-enterprise ethic in a predominantly capitalist economy, helping the individual is only undertaken as a means to helping all. This, at least, is the rhetoric.

The case for state intervention is multifaceted. In Britain today the scope for individuals to reduce their own exposure to flood hazards is strictly limited, although the individual farmer's rôle in agricultural drainage is quite crucial. Firstly, it is generally not technically possible or economic to provide flood alleviation for individual property owners or agricultural drainage for just one farm. Flood alleviation is an indivisible commodity. It affects other parts of the river system and also can be provided most efficiently in large quantities. Levées can be built around individual buildings, but community land-use control and overall river regulation or flood warning systems are generally more cost-effective and reliable. The water table of one agricultural area cannot be lowered unless neighbouring areas are affected: the hydrological system cannot easily be subdivided to coincide with property boundaries. Such hydrological interlinkages mean that collective rather than individual decisions are required. The state is invoked to seek agreement, comparable technical standards, and, consequently, the maximisation of social welfare rather than the possible anarchy of individual response.

A further reason for state intervention arises because in most communities only a minority are affected. The state's rôle, therefore, is to channel subsidy to this unfortunate minority. This argument for intervention is reinforced because, as with most natural disasters, flooding most affects those least resilient to its effects. Such people are typically those least capable of withstanding the financial impact of flood damage and least able to use technical arguments and political pressure to reduce their vulnerability. The impacts of flooding therefore tend to be selective, thereby widening inequalities between the rich and poor, the young and the old. Although there are obvious exceptions, those at risk in Britain from repeated flooding tend to be poorer than the national average. Flood alleviation can thus be redistributive by safeguarding the vulnerable at the expense of those already insulated from financial loss or the other adverse effects of flooding.

Irrespective, however, of such redistributive arguments a pervading attitude is that floods are natural 'Acts of God' and not the fault of those at risk. The state, therefore, intervenes on behalf of the unaffected community to rescue those afflicted who, as individuals, could not afford the protection they need. This is the 'welfare state' case for intervention: helping those who appear blameless for their own fate and thus, incidentally, helping to redress the 'natural' imbalances in an inegalitarian world.

A final reason put forward for state intervention in the land-drainage field is to promote increased agricultural production, subsidise investment and thus protect the profitability of farming. Such a reason perhaps contradicts the previous one – helping disadvantaged minorities – since those farmers who benefit most from subsidised investment tend to be those with the capital available to contribute their share. In any case, lowland farmers are amongst those most affluent in Britain. Nevertheless, increasing the profitability of the individual farm unit with investment in both new machinery and land improvement through drainage is seen by government as the means by which the collective good of increased food production can be attained. Main river arterial works are generally required, which the individual farmers perhaps could not finance and from which others concerned will benefit. The community, therefore, invests to seek a community-wide return. The consequent increased food production can reduce the cost of food imports and their adverse effect on the nation's balance of payments. The policy thereby seeks to enhance national – or European Community – self-sufficiency and security in a world where food is a scarce commodity carrying significant political power.

WELFARE PERSPECTIVES AND POLITICAL ANALYSIS

Flood hazards can cause misery and loss of life and may threaten personal health and family financial security. Flood damage to commercial, industrial and agricultural enterprises may mean that businesses become bankrupt and individuals may lose earnings or employment. Agricultural land drainage schemes may be necessary to maintain the economic viability of farming communities. The principal popularist 'welfare state' goals – prosperity, health, security and opportunity – are thus attainable only with the provision of a wide range of public services including flood alleviation and land drainage. Flooding puts these goals at risk so that avoiding floods is ultimately important to the maintenance and improvement of the quality of life. Wealth creation can lead to prosperity, but this is threatened by flood damage to property, agricultural crops or industrial production and by the economic losses thereby caused.

The problem of flood hazards and poor land drainage thus obviously lies principally in their economic effects, as properly analysed by Kates (1970) and others, and perhaps also in their consequences for personal health (Bennet 1970). But people are also affected by other adverse economic effects and other 'social evils' can also be relieved by public expenditure. Health can be improved by more investment in preventative medicine or on hospitals. Prosperity might be increased by higher levels of state investment in manufacturing industry. Agricultural production could be raised with greater investment in pesticides or advisory services. When seeking increased social welfare, therefore, why spend money on flood alleviation and land drainage? When evaluating the range of hazards people face, why worry about floods?

Thus analysed, the problem becomes one of deciding how much to spend to achieve the increases in social welfare that come from flood alleviation and land drainage in relation to other programmes for hazard reduction and other social welfare gains from hospitals, industrial investment and alternative means of agricultural intensification. This in turn involves decisions about the relative standards for the provision of a wide range of public services, each of which affects a different group of consumers. This is not just a technical matter concerning welfare economics or cost–benefit evaluations, although these may be involved. Such an analysis of the welfare products of flood alleviation and land drainage cannot be divorced from the question of 'welfare for whom?', particularly because flood alleviation and land drainage affect only a minority but contributions to expenditure come from a wider public.

Since the aim of flood alleviation and land drainage is to increase social welfare and since the state has, for the reasons discussed, taken a rôle in this process of resource allocation, then decisions should clearly be collective and political in nature rather than just involving individuals. Those making decisions concerning standards of land drainage and flood alleviation need to recognise that different interests or 'communities' will have different views on what is beneficial. Such decisions are inseparable from the overall debate about total welfare state provision. This in turn focuses attention on the institutional context in which decisions are made because here is the locus of political power.

In Britain, this context focuses our attention principally upon the state agencies affected and the community groups seeking to use their economic power and political bargaining to maximise their share of the nation's wealth and resources. Such a focus coincides exactly with the clear understanding most land drainage engineers have of the political influences on decisions with which they are concerned, and the significance of power groups within this political process. The politics involved are generally not explicitly party political in basis, but this in no way reduces their importance.

Attention is also focused, with this perspective, on the consequential effects of expenditure on flood alleviation and land drainage: within finite and tightly controlled public expenditure ceilings more money spent here means less for hospitals or motorways. All these effects cannot be discussed here, but we should note that the interlinkages inherent in the hydrological cycle are mirrored in the parallel economic and political structures within Britain, where each decision on resource allocation has an opportunity cost in terms of alternatives forgone. Flood alleviation and land drainage cannot be seen, therefore, just as 'a good thing'. In addition to unfortunate environmental effects or other adverse impacts it may also mean marginally lower standards of public service provision elsewhere.

OTHER RESEARCH PERSPECTIVES

The 'welfare' perspective on flood alleviation and land drainage discussed above contrasts markedly with the philosophy apparently underlying much of the previous research on flood hazards. Most of this research, largely emanating from the USA and Canada, has stressed the rôle of the individual in reducing their vulnerability to hazards, following the mould of enquiry established by White (1945).

White's analysis was developed in recognition of the demonstrable failure of the collective approach to flood alleviation followed during the 1930s to 1960s in the USA. He replaced an emphasis on narrow approaches to collective action, embodied in large-scale structural flood-control schemes, with a broader concept of collective action termed 'floodplain management'. This emphasised evaluating the full range of possible adjustments to reduce the adverse effects of floods and the need to persuade the individual to take action. One result, the large-scale Federal Insurance Program initiated in the late 1960s, encouraged individual communities towards local regulation of land use by providing the incentive of subsidised flood insurance. The overall approach emphasises influencing people rather than flood water and focuses on a local community framework rather than on Federal construction agencies. The ethos is one of self-help, drawing upon the American philosophy of individualism, and dominated by a tradition of unconstrained land use which had allowed unfettered encroachment of flood-prone areas.

In the early Chicago research the rôle of collective decision making, government action, institutional analysis and the politics of decision making were not emphasised. There has thus been little analysis of flood alleviation in relation to other state services – just in relation to other hazards – and consequently little debate about standards of flood alleviation. This partly arises because the researchers' behavioural framework of analysis, and thus the problem posed in terms of individuals making suboptimal decisions, leads implicitly to an acceptance of total flood protection that the public unthinkingly demands.

The American research framework thus has its roots in the American self-help, free enterprise, capitalist philosophy. It is not inapplicable elsewhere, but the 'welfare state' philosophy – or rhetoric? – in Britain forces a different perspective which gives more emphasis to collective decision making, institutions and political analysis. Such emphases, moreover, are not without their parallels in recent research from the USA and elsewhere (Platt & McMullen 1979, Platt 1980). However, in Britain this alternative perspective is coincidentally helped by the links between agricultural drainage and flood alleviation. These highlight the political nature of decision making by demonstrating the decisive rôle of one particular power group, in this case the 'farming lobby'.

A theoretical context

It is useful to remind ourselves that we need and use theories to assist our understanding of the real world. A theory or theoretical structure is a set of scientific laws derived either inductively from generalisation about unordered facts or deductively from logical argument and reason. Such a theory will enable us to deduce sets of hypotheses which may be tested against empirical data collected for that purpose; the data themselves are not absolute or neutral but inevitably reflect our ideas and values. The more hypotheses we can check in this way the more confident we can feel in the validity of the theory and use it to predict future events and to explain what we observe (Harvey 1969, p. 35). Theories thus enable generalisation, explanation and prediction so that science – used either consciously or unconsciously – provides us with a consistent, coherent and empirically justified body of information on which to base our understanding of the world.

Many theories are involved in different parts of the complex field of study that encompasses planning and decision making about floods and land drainage. These include economic theory, hydrological theories, theories about public participation, decision making and consultation, and ecological theories about successional or sudden change with the disturbance of ecosystems following drainage. We are thus simultaneously involved in theories of natural and physical science and those from the social sciences and the hybrid areas of knowledge that deal with the interaction between society and environment. Most of these theories cannot receive adequate attention here but two overriding areas of generalisation warrant consideration since they illuminate our observations and suggest further hypotheses for investigation. These are *hazard–response theory* and *decision-making theory*. Neither is particularly strong, in the sense that the laws of thermodynamics are strong, but this area of enquiry is young and unequivocal laws to guide policy analysis are sparse indeed. We cannot ignore, however, the fact that we are not handling our observations in a vacuum but are working with and towards theories or generalisations.

HAZARD–RESPONSE THEORY

Within the social sciences the analysis of natural hazards has attracted researchers from many disciplines including geographers, economists, psychologists and anthropologists (e.g. Burton & Kates 1964, Mileti 1975, Janis & Mann 1977, Torry 1979, Quarentelli 1980, Sorkin 1982). During the 1960s and 1970s hazard–response theory developed inductively as generalisations accumulated from case studies, predominantly of floods (White 1961, Kates 1962, 1970, Burton *et al.* 1968, 1978).

Hazard–response research has sought to clarify the extent and reasons for human occupation of hazard zones, how people perceive extreme events, the range of hazard-reducing adjustments and the process of selecting hazard-reducing options. Kates (1970, p. 2) suggests that with this research we begin 'to structure a primitive general framework of human adjustment to natural hazards in which we try to preserve its human ecological perspective'. Such a framework or model of reality sees interactions between people and nature in the short run as a stable self-regulating dynamic equilibrium. In the longer time period people become increasingly dominant over their environment, although little is known of long-run adaptive processes.

At its most generalised, hazard–response theory is part of a wider analysis of natural resources in the vein pioneered by Zimmerman (1951) and others. At its most detailed, the theory is concerned with the mechanisms of decision making within the process of adaptation whereby people and society as a whole aim to 'seek in nature what is useful and attempt(s) to buffer what is harmful' (Kates 1970, p. 1). These two aspects are discussed below in turn.

Hazard–resource relationships Neither hazards nor resources are absolute phenomena: they result from *relationships*. Natural *hazards* have been viewed as the detrimental consequences of peoples' use of their environment; beneficial outcomes of environmental use are labelled natural *resources* (Kates 1970, Burton *et al.* 1978, p. 20). According to Zimmerman's (1951) complementary theory, natural resources are the product of interaction of people, culture and nature. Hazards are 'resistances' or barriers that interfere with resource exploitation. Resources are subjective phenomena and their value is dynamic and relative to the changing wants and needs of society and the technology at its disposal, rather than absolute and static; the same applies to natural hazards.

In our case both floods and waterlogged soils are environmental characteristics that give rise to hazards either in the form of flood losses or as low soil productivity due to poor drainage (Fig. 1.1). On the other hand, both flood-prone areas and wet agricultural land provide the community with resource potential by embodying exploitation opportunities, both urban and agricultural. These opportunities cannot be realised, however, without investment in flood alleviation and land drainage. Such investment is only useful if the resources created are needed and perceived as such.

However, the process of modifying environmental relationships to increase resources or generate wealth is complex. This, firstly, is because of the opportunities and dangers inherent in disturbing a complex system (Fig. 1.2). We have postulated, for example, how extensive agricultural land drainage might exacerbate downstream urban flooding by altering the timing

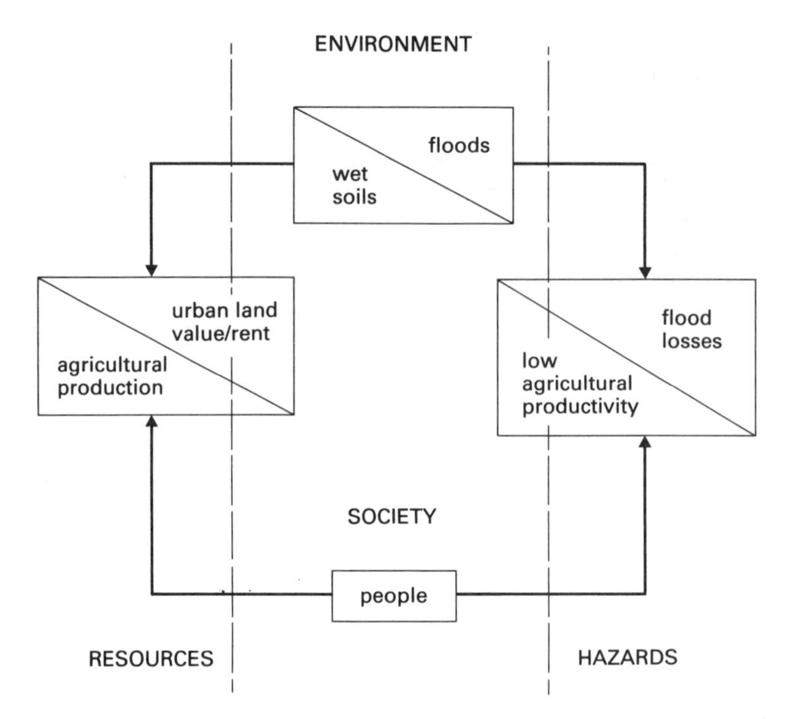

Figure 1.1 'Resources' and 'hazards' are terms used to describe relationships between society and the environment. These relationships create either resources, comprising food produced from agricultural land, or hazards in the form of flood losses.

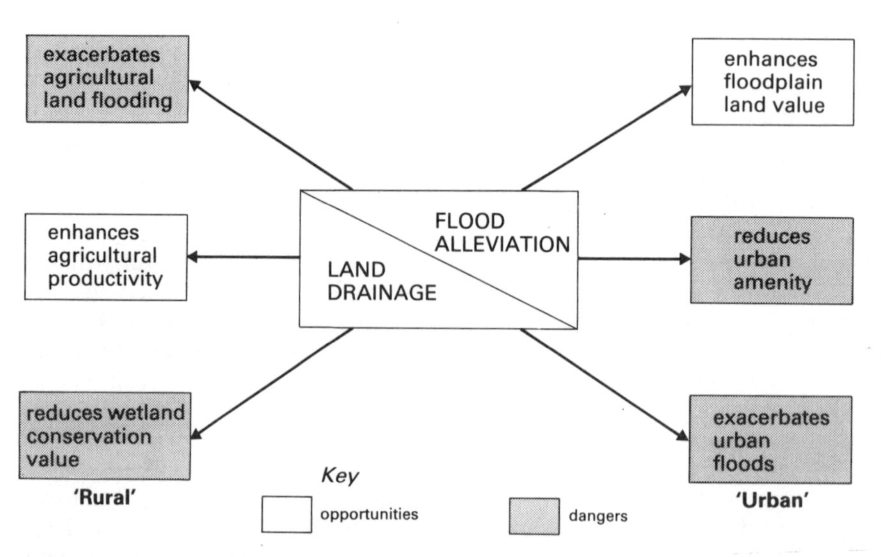

Figure 1.2 Opportunities and dangers from flood alleviation and land drainage.

of runoff. Other dangers are less tangible but no less significant for that. Secondly, since resources and hazards result from relationships, one person's resource may be another person's hazard. Agriculturally unproductive wetlands may be of value to conservationists because of their scarce flora and fauna or their special landscape features. Drainage to increase agricultural productivity and farm incomes may reduce these conservation and landscape values. Urban flood alleviation schemes to reduce industrialists' flood losses might also reduce the amenity value for local residents of highly prized riverside locations. The adaptive action taken to enhance resource exploitation may create or worsen hazards elsewhere and may disadvantage certain groups while benefiting others. In practical terms the outcomes may well be difficult to determine because of the complex interlinkages between hazard reduction and resource enhancement or diminution.

In summary, several points emerge from this analysis which complement previous discussions. First, the simple trade-off between hazards and resources is more complex than Kates (1970) suggests: simply reducing hazards does not necessarily produce resources. Secondly, the nature of resources – and hazards – is relative: increased food production or reduced flood losses may be seen by some as a good thing but viewed by others with indifference or even hostility. Thirdly, any actions to adapt the environment for society's wants and needs may have unforeseen adverse effects elsewhere or may simply be giving resources to one section of society to the detriment of another. Fourthly, a measure for the success of flood alleviation and land drainage must be the degree to which the adverse consequences of these environmental modifications can be anticipated, minimised or avoided to yield a net increase in resources in the form of increased social welfare: no small task!

The adaptation process The process of adaptation or adjustment to hazardous environments is central to hazard–response theory (Kates 1970). An apparent deficiency of this theory, for British circumstances at least, is its over-emphasis upon adaptation by the individual (Fig. 1.3).

Although never fully tested, Kates's model seeks to explain how individuals adjust to natural hazards by applying the now classical steps in decision-making theory (Simon 1957). Kates postulates that for each occupant of flood-prone areas there is a perception threshold below which they do nothing. When the threshold is reached the person begins to seek mitigating adjustments (Fig. 1.4), and evaluates them for the appropriateness to local circumstances, their technical feasibility and their economic worthwhileness. The threshold is related to the person's experience of past hazards and specific personality characteristics that include attitudes towards fate, the tolerance of conflict between the ideal and the achievable and their propensity to gamble. Individual adoption of adjustments is related primarily to experience, wealth, the capacity to undertake adjustments and their importance to livelihood; to a lesser degree individual variation and personality are significant (Burton *et al.* 1978, p. 146).

The adaptive behaviour of individuals is boundedly rational (Simon 1957). Thus one limit to rational decisions may be the lack of relevant

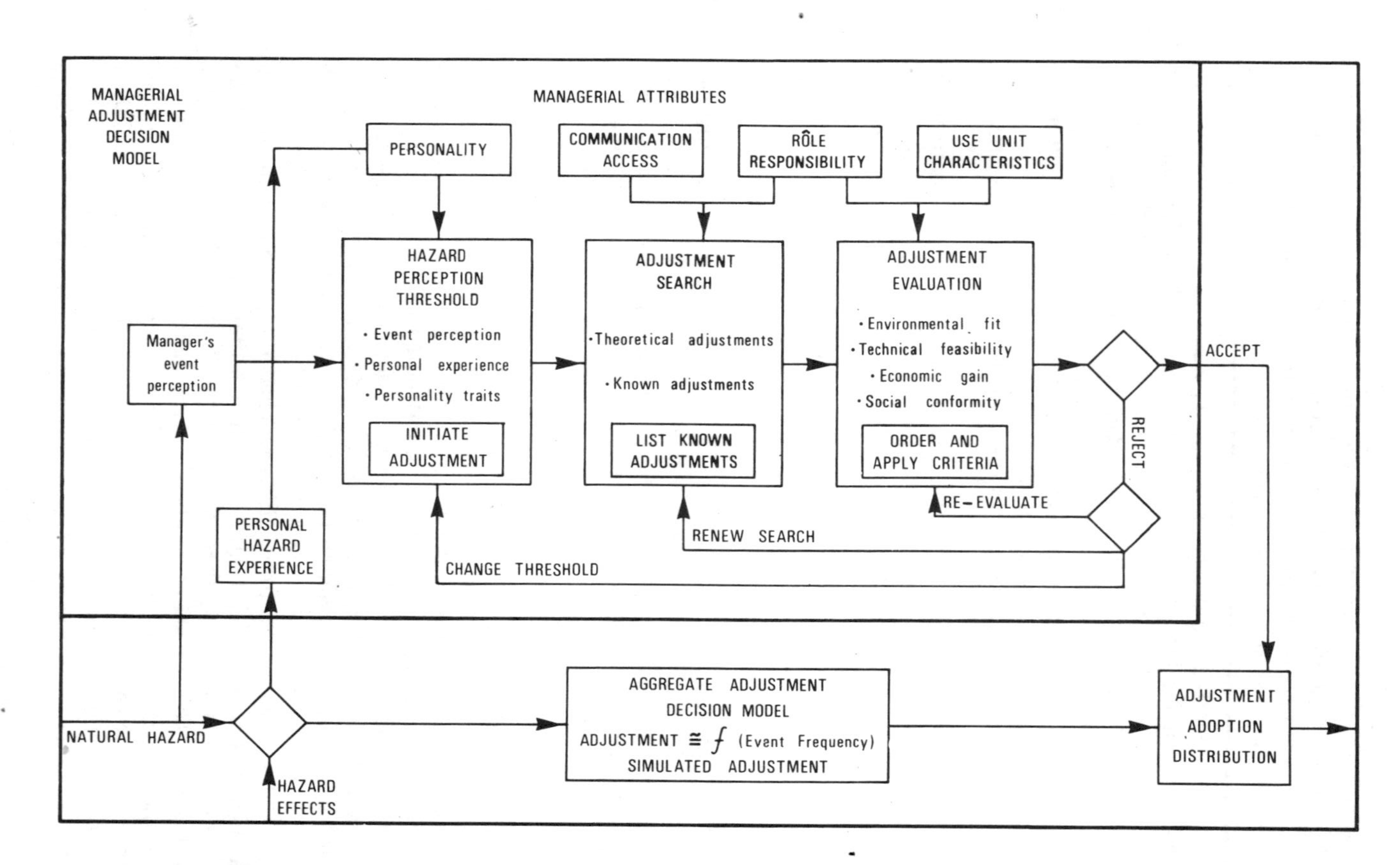

Figure 1.3 A conceptual model of the adjustment process (from Kates 1970).

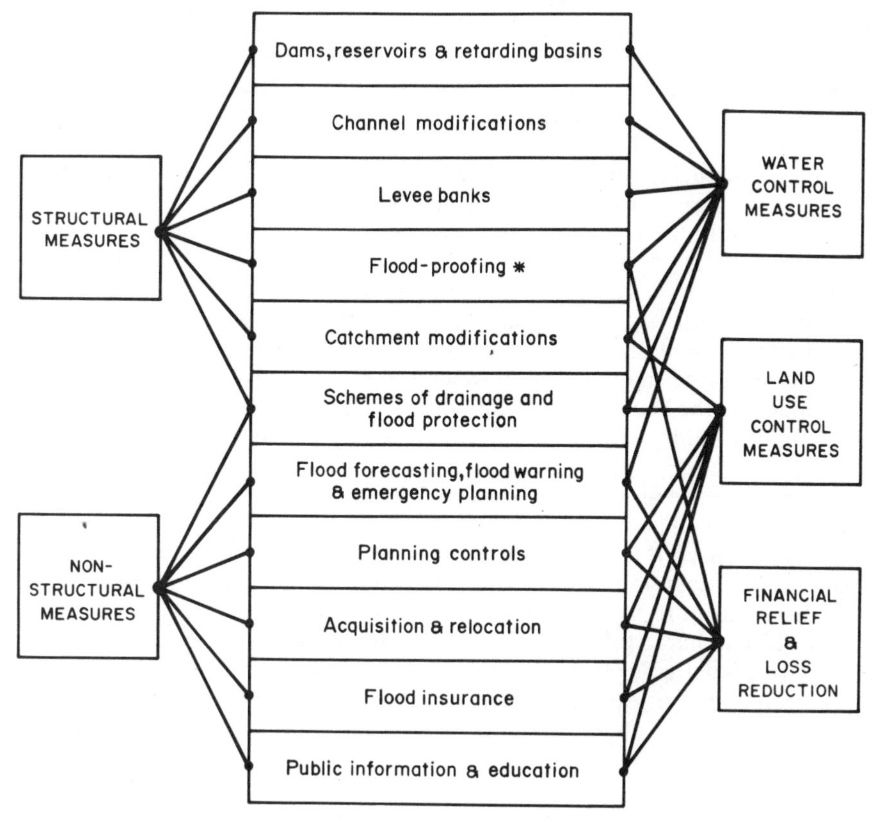

Figure 1.4 Available measures to mitigate flooding and its effects (from Victorian Water Resources Council 1978).

information. People also have difficulty in drawing valid conclusions about random events, especially if their experience is limited. They commonly attempt to reduce uncertainty by attributing regularity to flood events or by denying the existence of flood risk. Evaluation of alternative actions is simplified to a manageable 'ordered choice' process where individuals select among a very small number of alternatives rather than simultaneously analysing a wide range of options (Burton *et al.* 1978).

Community level adjustments are modelled in the crudest of forms by Kates (1970): adjustment is merely a function of event frequency (Fig. 1.3). Thus very few adjustments can be expected where the probability of floods is low and where the flood-prone population is relatively clear in their perceptions that such events will not occur. A large number of people will adopt mitigating adjustments where the probability of flood occurrence is high and where the population is also clear that floods will occur in the future. Between these extremes of relative certainty there is an intermediate probability of flood occurrence and the greater uncertainty is reflected in high variability in the adoption of adjustments by individuals. Kates thus models community adjustment simply in terms of the aggregate actions of individuals.

This oversimplification of the community adaptation process is corrected to some extent in later work. Burton *et al.* (1978) explain briefly the rôle of authorities in providing laws, regulations and incentives to adapt, in deploying resources and technological adjustments, and in dispensing hazard adjustment services. They also recognise the importance of special interest groups in influencing adjustment decisions and the rôle of the official bent on maintaining and extending his or her organisation.

However, rather than attempting to provide an explanation for, or even an analysis of, this decision-making process at the community level – which is complicated in their case by the multicultural context – Burton *et al.*'s main thrust is still just to evaluate the range of collective adjustments and describe the possibilities for national policy. In this respect they appear constrained by the aims of the paradigm in which they are working which are to increase the range of adjustments away from structural hazard control rather than to understand fully the processes by which decisions are made.

Community hazard-reduction decision making An advance on existing hazard–response theory is possible by identifying more carefully the forces that influence hazard adaptation decisions, at least in Britain. Later chapters expand our understanding of these forces and influences and demonstrate how they contribute to decision making.

Some researchers (O'Riordan 1971, p. 202; 1981, p. 246; Sewell 1973) attempt to model collective decision pathways by identifying and generalising the sequences of events leading to decisions. Others complain that 'the process of collective choice appears erratic, close up (secretive), it appears

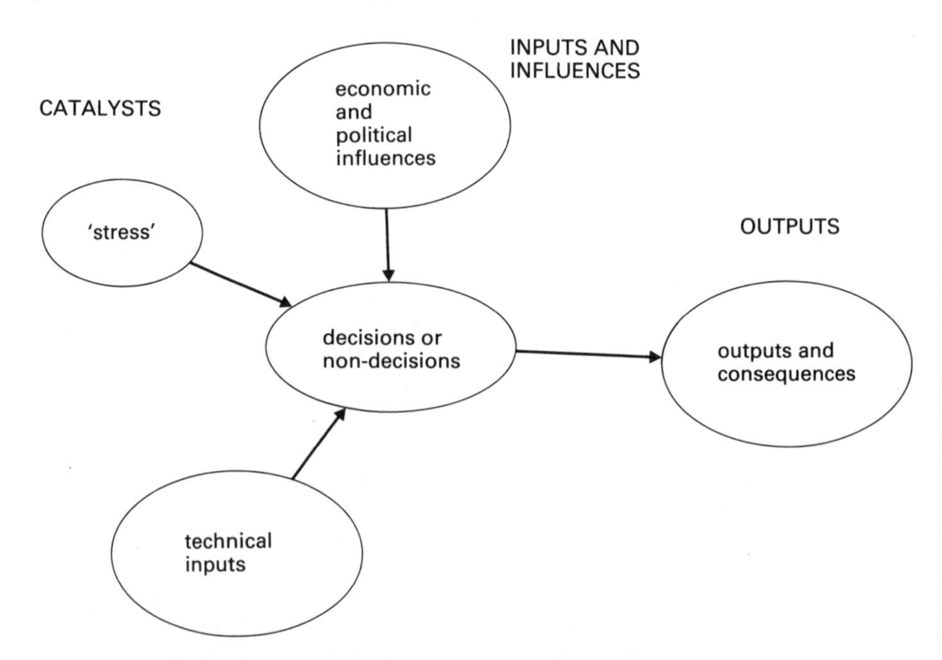

Figure 1.5 The basis of the 'community' or political decision process.

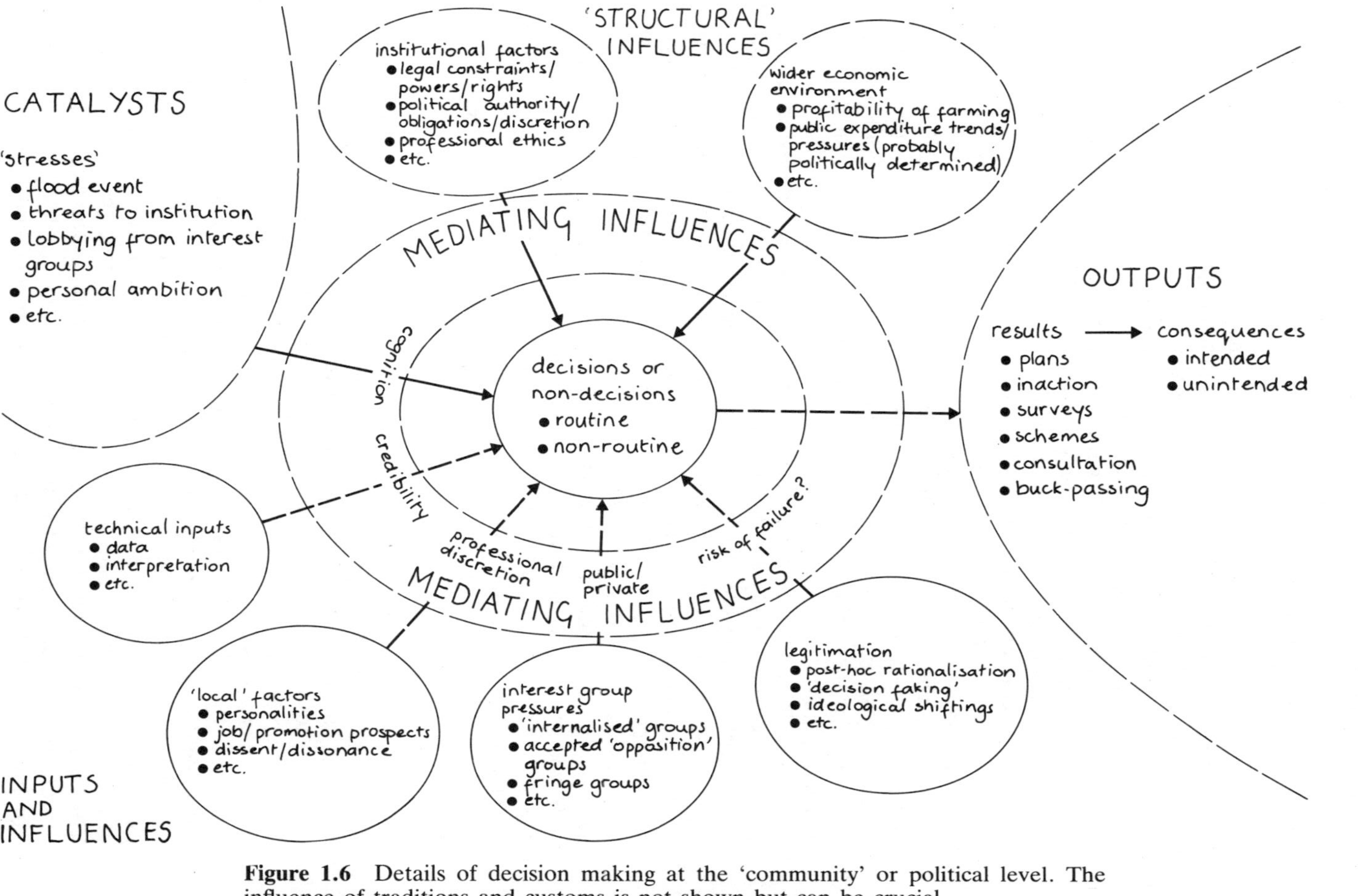

Figure 1.6 Details of decision making at the 'community' or political level. The influence of traditions and customs is not shown but can be crucial.

complicated by the unique and situational, haphazard and unpredictable' (Burton *et al.* 1978, p. 147). Even O'Riordan (1976, p. 65) suggests that 'there are too many special factors involved to make important generalisations'. He emphasises instead the particularities of key decision makers in conditioning policy making.

We prefer not to describe a single or small number of decision routes since in reality decisions are arrived at in innumerable ways. Generalisation of decision sequences beyond what is obvious is not possible. In simplistic terms we view decisions as arising from a variety of inputs and influences which produce a variety of outputs and consequences (Fig. 1.5). However, in contrast to Kates's (1970) hazard–response model and O'Riordan's emphasis, with their concentration upon the rôle of individuals, we prefer to emphasise the influence of economic and political forces upon decisions (Fig. 1.6). Our experience indicates that institutions and macroeconomic forces are more important than individuals in Britain in shaping decisions and, moreover, that institutions are more than just collections of individuals. The schema in Figure 1.6 does not and cannot *explain* individual decisions, but it *describes* comprehensively the forces operating on those making decisions. As such it can be used to ask questions about particular cases rather than pinpoint how and why decisions were made along a particular pathway of decision making.

Figures 1.5 and 1.6 show that the need to make decisions arises from *stress* (Kasperson 1969) which may take many forms. These include a flood event with its consequent damage and anxiety, or flagging agricultural profitability related to undrained soils. They may include the political or professional ambition of someone who is influential, or threats to the survival of drainage agencies and the employment of those within them. Decisions can thus be initiated in very different ways and not, as modelled by Kates (1970), simply by the environmental hazard itself.

Nevertheless – as in Kates's model – for decision making to begin, to counter this stress, someone or some group of people must perceive a need or incentive to act. Each 'actor' involved is subject to a variety of *pressures* both from the original stress factor and from the forces affecting the decisions they may make. These pressures provide the motivation for action. Farmers, for example, are usually concerned with maintaining or increasing their profits from the land. Members of the public affected by flooding are usually concerned to protect their property from future floods. Land drainage engineers have at least one eye upon preserving their jobs or seeking advancement. 'Conservationists' will be motivated to protect wetland areas from what they see as threats from drainage.

However, as shown in Figure 1.6, the effects of these pressures are *mediated* by other factors rather than universally and mechanistically affecting decisions. Flood alleviation decisions are thus undoubtedly affected by cognition or perception of risk. The academic or professional credibility of technical inputs affects their adoption. The extent of professional discretion to depart from set procedures will affect the influence local personalities will have on policy making. The need for rationalisation or legitimation of decisions, to protect the decision maker, appears to be conditioned by the risk of failure inherent in the decision being made.

Once a group of those involved in these decisions has at least one interest in common, they may begin to operate as an interest group despite other issues dividing them. They will organise to pursue their ends with various degrees of activity or passivity, using both public and private consultation and bargaining with others involved. The range of such groups involved in British flood alleviation and land drainage is large, and each is subject to different pressures, reflecting their basic needs and wants.

However, the nature of the pressures on those making decisions is affected by the wider environment in which they operate and decisions cannot be understood simply by understanding pressure-group politics. Farmers and property developers seeking to intensify the use of floodplain land thus react to the potential for higher profits within a wider and often contradictory economic environment of constantly changing economic fortunes. Their propensity to invest is, therefore, conditioned by the prevailing levels of return on such investment in relation to other investment opportunities. The reaction of conservationists to drainage proposals also cannot be separated from the prevailing trends in the depletion or otherwise of nature conservation resources. The land-drainage engineer and authority, in Britain at least, is profoundly affected by trends in public expenditure and the underlying political philosophy creating these trends. Therefore, to treat the individual decision or circumstance in isolation from its context is dangerously naïve, and to separate pressure groups from the overall place in society of their members is to lose sight of their ultimate power base.

Whether a person, group or drainage agency decides to adapt to stress depends partly upon their *legal* position, and the *political authority* this brings, within 'the political culture which frames all policy evaluation and execution' (O'Riordan 1976, p. 229). Individuals have drainage rights and responsibilities in law. Drainage agencies are given responsibilities, duties and powers by legislation and must act within these constraints. Thus the law acts as a guide as to what can or cannot be achieved, and relates ultimately to the interests of those with power, but individuals and agencies have some areas of discretion often demarcated by precedence or professional ethics.

In this way 'local factors' working within drainage agencies and elsewhere may also affect decisions. For example, the personality, professionalism and ambition of leading officers within the agencies can significantly affect the promotion or non-promotion of flood alleviation or land drainage schemes. The degree of local unity or dissonance will influence the efficiency with which plans are prepared and executed. It will also affect the political credibility of that local unit within a wider bargaining structure of official or other organisations. The propensity of officials or others to risk the failure of their plans will affect the type of plans prepared. The extent of 'profit maximisation' can be important: individual farmers may not be interested in draining land to make their enterprises more efficient, and thus increase their incomes, despite an overall trend in this direction. An individual administrator's experience also has a bearing upon decisions, those with most experience tending to promote caution and conservatism. Local needs to preserve jobs by ensuring continuity of work can influence organisations by encouraging delay or even leading to the promotion of less than worthwhile or spurious schemes.

Finally there are the *technical inputs* which affect decisions and are themselves affected by their professional or general credibility. The design of flood-alleviation and land drainage schemes is assessed in terms of their hydrological feasibility and hydraulic capacity. Hydrological data and analyses are, therefore, important inputs into any decision. The government increasingly emphasises the need for economic efficiency and thus for obtaining value for public expenditure. Economic inputs to the appraisals of schemes, therefore, also assume some importance. Most drainage schemes – whether urban or agricultural – have some environmental or amenity impact and some form of environmental impact assessment should also form a further technical data input to decision making.

The significance of these technical inputs is not just their absolute accuracy, or indeed their relevance, but the way that they facilitate design, legitimate predetermined designs, or provide counter-arguments for the opponents of those 'making the running' in the decision process. Indeed these data and techniques themselves are resources; their value is relative to the positions of those using them, and withholding information, for example, can significantly enhance its importance. The information and techniques are not neutral or value-free, but reflect judgements by those concerned as to what is important or problematic.

As shown in Figure 1.6 all these pressures and inputs lead to decisions on action, or decisions not to take action, or decisions not to take decisions! Each will have different consequences, some of which may be unintended. Each consequence – if possible including the unintended – will need to be rationalised or legitimated by those making (or not making) decisions. Plans will only be adopted if they can be justified in terms acceptable to those in higher authority who might question their wisdom. Many outputs will, however, be rationalised after a decision has already been made on other grounds. Thus decisions taken for the 'wrong' reasons – perhaps political expediency – will nevertheless need to be justified as being worthwhile, professionally sound or publicly acclaimed. Even the 'right' decisions may well need to be 'sold' to a sceptical public and those providing finance. Thus the need for legitimation inevitably conditions both what is decided and much of the behaviour of those making the decisions.

DECISION-MAKING AND PLANNING THEORY

Although the community decision-making schema discussed above provides some insight into the forces that influence decisions concerning flood alleviation and land drainage, it is less successful in explaining how decisions are actually made. In this respect some other models may be more useful.

Decision-making research has developed primarily within political science, planning and public administration (Lindblom 1959, Dror 1964, McLoughlin 1969, Chadwick 1971, Faludi 1973). Attempts are being made with this research to render those subject areas more intellectually rigorous – and thus respectable – by constructing a theoretical and, therefore, a 'scientific' approach to decision making. One model is essentially a descriptive generalisation. This characterises decision making and planning as 'disjointed incrementalism' whereby decisions are made hurriedly in response to crisis through a process of 'muddling through' (Lindblom 1959). Goals are not discussed, alternatives are narrowly defined and poorly evaluated, and ends are adjusted to suit practicable means. A good policy or plan is merely one on which agreement is reached and this agreement is more important than the substance agreed upon: decision making in itself becomes central and the process is one of continuous marginal policy adjustments as the environment of the decision maker shifts. Radical change cannot be accommodated and planning becomes the ultimate in pragmatism. Such a position by decision makers is supported or legitimated by a pluralist conception of society in which groups compete for available resources and the most powerful groups are 'right' and therefore win but never dominate, and everyone eventually gets a share. Those making key decisions thus must retain flexibility and act incrementally to ensure that they can accommodate the changing pressures from a dynamic interaction of many competing interests.

A second model is more prescriptive or normative in suggesting an 'ideal' planning process that adopts a rational or 'linear-deductive' approach, perhaps having some parallels in Kates's (1970) generalised or ideal process of adjustment adoption. This linear-deductive approach (Fig. 1.7) initiates

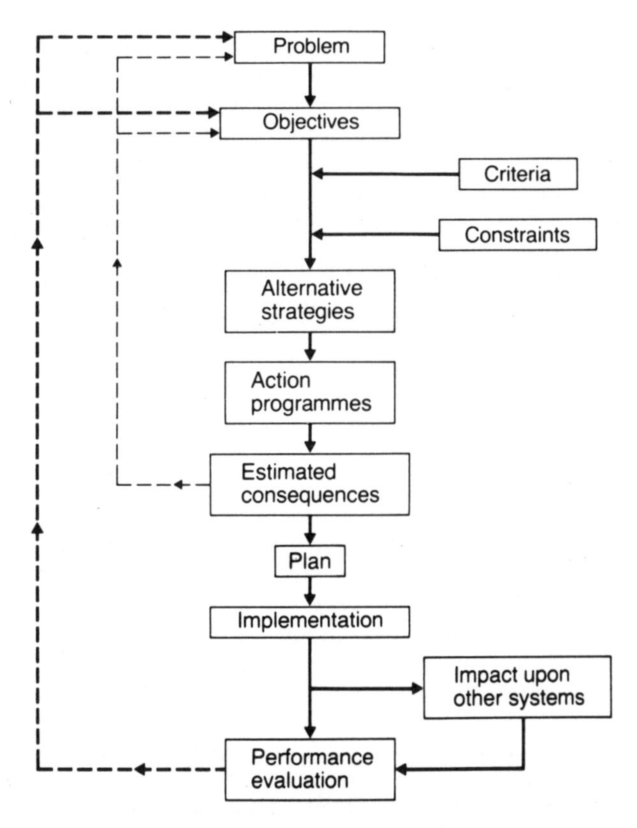

Figure 1.7 One interpretation of a linear deductive decision sequence (from Mitchell 1971).

decision making with a clean slate – back to fundamentals each time – and proceeds through a predetermined or 'linear' sequence of steps within which each stage must follow directly and logically from the stage preceding it. The sequence begins with a decision to plan and establish goals or problem identification. It proceeds through the examination of possible plans or schemes, the evaluation of alternatives, and ends with plan implementation and the monitoring of policy effectiveness (Faludi 1973). Such a process requires a consensus on the rules leading from one stage to another – for example, deriving objectives from goals – and that evaluation should be unambiguous to all concerned; goals themselves must also be non-contradictory. The process of decision making should leave the planner or engineer as a detached, neutral and objective commentator rather than someone feverishly involved with the issues and committed to particular outcomes.

Criticisms of this 'ideal' decision-making model are formidable (Simmie 1974, Ch. 6). They include arguments that the model is impossible to fulfil both in terms of the rigours inherent in its logic and the more immediate constraints of time, information and other requirements. Also as a technique of making decisions its mechanistic approach is alien to the methods of political action. It denies that planners and engineers operate within systems with political constraints. The process can also impose a consensus where none exists owing to the relentless process of stage-by-stage decision sequences forcing agreement where disagreement remains real. In addition this conception of decision making tends, by its concern for constant evaluation of alternatives, towards a positivistic emphasis on data rather than political debate about means and ends. This is because data collection and analysis is seen as politically neutral and emphasis on this aspect of the 'problem' supports the supposed neutrality of the whole process. Endless discussion of goals and means, on the other hand, cannot be non-political and as such does not square with the supposed scientific nature of decision making.

Etzioni's (1967) intermediate model of decision making has some following. Here policy making features a series of incremental decisions but policy makers are also cursorily scanning more radical alternatives which are adopted periodically. Choosing alternatives to scan is based on some prior ordering of priorities, rather than pure pragmatism, but decision making is not so systematic that set criteria are used such as the maximisation of net economic gains.

Gutch's (1972) explanation of this model notes that both linear-deductive and disjointed incrementalist processes deal inadequately with the difference between plan design and its justification. These, therefore, are separated by Etzioni within a cyclical model, also pioneered by Boyce *et al.* (1970), which proposes that the plan design process starts at any stage rather than just with goals – it can begin with a survey or a choice of plan type – and then proceeds along a set path of prescribed steps. The goals are deduced from the plan: they are isolated 'at the end' and subjected to an independent justification process or post-rationalisation. As Hamnett (1973, p. 23) indicates 'the main feature of this approach is that it acknowledges that goals cannot be formulated independently of plans in practice' and that 'the political awareness of the incremental approach is . . . retained' with 'a useful degree of "ad hocery" '. An outward seal of legitimacy is bestowed by an apparently systematic planning system, while retaining the rôle of professional judgement rather than relegating planning to a mechanistic linear process.

These models of decision making have some use in identifying and categorising planning systems and perhaps in predicting planning processes. However, in many respects they are abstract and reduce planning merely to decision taking. Flood alleviation and land drainage are fundamentally concerned, however, with allocating scarce resources for welfare needs (Penning-Rowsell & Parker 1983b). It is thus inseparable from other areas of public expenditure for hospitals, schools and other areas of water planning. Each involves bargaining between groups for these resources within the context of Britain's advanced capitalist economy. Inseparable from this bargaining process are the pressures within British society to accumulate wealth and to protect private property. Expenditure on agricultural improvement reflects the first of these pressures, and reduction of flood damage reflects the second.

We can therefore see that in this process of making decisions there are 'losers' and 'gainers', as there must be with any planning. Various sectional interests clearly have a dominant influence upon flood alleviation and land drainage – principally the multifarious farming and landowning interests – and these interests are supported by government policies and subsidies. This should not be taken to imply that land drainage engineers deliberately seek to promote the interests of these groups, but that the processes of government and the forces within our economy necessarily support the wealth-generating capital-owning classes upon which the economy of the country is seen to depend. Thus our analysis of flood alleviation and land drainage cannot just use the 'bench marks' of the hazard–response paradigm, or theories of policy evolution or decision making. We must also seek to relate policies and plans in this specialist field to the wider political economy of the country in which it occurs.

Problems

In the chapters that follow we present a comprehensive analysis of current British flood alleviation problems and practice, excluding engineering design which is not our domain. We also develop our understanding of decision making. In doing so we follow the pattern set out in our community or political decision-making schema (Fig. 1.6) and we amplify the varied influences on decision making it identifies.

Some of the overriding problems in any planning – and the analysis of planning – concern uncertainty and imperfect knowledge. The chapters that follow therefore describe and evaluate the available data sources, analytical methods and the many problems in their application. These problems require at least three kinds of response. First, a working knowledge is required of the latest techniques of measurement and modelling of hydrological, hydraulic, economic and environmental impact parameters.

Secondly, there must be a clear awareness of the imprecision associated with these techniques, as well as a continual search for improved data and methods. In this context we suggest some possible improvements. Thirdly, river engineers and planners increasingly require a demandingly wide range of knowledge of the implications of their work. This includes a keen awareness of institutional arrangements and political processes. Since the tendency of all professionals – engineers, researchers and planners alike – is to focus inwards on their speciality, developing the ability to comprehend the 'externalities' of their work is one of the most difficult and challenging of tasks.

Before discussing any of these matters in Chapters 3 to 6, however, Chapter 2 elaborates the institutional context of British flood alleviation and land drainage which dominates the schema in Figure 1.6. Our analysis of legal powers, institutional structures and finance thus focuses on the way these may influence the decisions concerning agricultural drainage and flood alleviation. The final chapter then re-examines the relationships between individuals and institutions and thus reviews the insights provided by hazard–response and decision-making theories for flood alleviation and land drainage in Britain. We also highlight key areas for further research and the implications of British experience for both the future development of theory and for flood hazard reduction and drainage policies elsewhere in the world.

A developing scene

Land drainage and flood alleviation policy in Britain continues to evolve, mostly very slowly but sometimes by leaps and bounds. Inevitably this means that some of the detail in the ensuing chapters will become out of date, but the over-riding framework – stressing the wider societal context to individual policies – will remain valid. It will be a test for the reader to relate the empirical circumstances of their day to this context and, hopefully, thereby add to the depth of our policy analysis.

The events of 1984 to 1986 illustrate this evolution. Growing EEC agricultural surpluses led the government to reduce grant levels for infrastructure investment including agricultural drainage (Ch. 2). This move came despite the traditional political support from agricultural interests for Conservative governments (but then modern Conservatism is not land based). A simultaneous and perhaps deliberately complementary revision of MAFF guidelines for agricultural benefit assessment acknowledged the conservationists' arguments against biased cost–benefit analysis (Ch. 4), supported by Treasury scepticism of this investment and the government's desire to reduce public spending at virtually any political cost. The environmental impact of drainage schemes has become more widely acknowledged (Ch. 5) and indeed conservation is now being promoted, not least to protect river engineers' jobs but also perhaps as a government rationalisation of expenditure cut-backs. 'Compensation' at £50 per acre per year (£122.50 per hectare) to Halvergate farmers to maintain the drainage *status quo* has been a major innovation following a lengthy battle (Ch. 6).

The result is perhaps a stay of execution for an outdated landscape rather than a total conservation victory: the farmers win either way.

The power of the Ministry of Agriculture, Fisheries and Food and its supporters continues, however (Ch. 2). A 'green' consultation paper in 1984 (Cmnd 9449), on the future arrangement of land drainage, aired some radical possibilities, such as removing the special position of land drainage and flood alleviation within Water Authorities (Ch. 2), and a complete overhaul of financial arrangements on a more stringent 'beneficiaries pay' basis (Ch. 4). What was not addressed was the central rôle of the ministry itself. However, on past performance, the system will escape largely unscathed although Water Authority privatisation and the 'rationalisation' of water authority staff poses a more serious threat. One ministry answer was to move more forthrightly into coastal protection and sea defence, on the back of which much traditional drainage work could be funded, and this responsibility was taken by the Ministry from the Department of the Environment in 1985.

Most officials and water authority engineers are eagerly awaiting another 1947, 1953 or 1968 flood season to restore their political positions, shaken by a government which reflects the growing dominance of finance capitalism rather than landed interests. To individuals, the mid-1980s will appear as a turbulent time of radical change. Future historians looking back are more likely to see a number of minor inflexions on the overall pattern of continuity in British drainage policy reflecting the changing economic fortunes of Britain in the world.

2 *The institutional context*

The institutional perspective

Engineering, farming and floodplain planning are all activities that occur in Britain, as elsewhere, within an institutional context. This context profoundly affects their overall policies and trends and their day-to-day operation. We therefore need to understand the components of this institutional context, including the driving forces and 'laws of motion' behind the institutions and organisations involved. This will help us to explain their policies and activities and perhaps predict their future behaviour. In this respect the institutional perspective developed in this chapter both uses and elaborates on our conception of community decision making, with its emphasis on the economic and political forces affecting decisions (Ch. 1, Fig. 1.6).

INSTITUTIONS, INDIVIDUALS AND ORGANISATIONS

The broad definition of the institutional context we use resembles that presented by Howe (1977) who identifies three main components. These comprise laws, organisational structures and publicly held values and perceptions. Together these form a 'shell' within which technical planning and political bargaining take place. In our definition the institutional context also includes the rôle and institutionalisation of the individual, in our case as a riparian owner or farmer.

However, a necessary contrast was drawn in Chapter 1 between hazard–response theory, with its emphasis on the rôle of the individual, and community hazard–reduction decision making. This contrast suggested the naïvety of interpreting the individual's decisions in isolation from a wider analysis of the administrative, economic and political forces affecting decisions. Nevertheless the significance of individuals' cognition and adaptation to hazards must not be ignored. Indeed, as we shall see, British land drainage law bestows important powers and responsibilities upon certain individuals and the individual has an evolving rôle which is itself 'institutionalised'. Moreover, many hazard reduction and agricultural improvement policies in Britain, although dominated by institutional factors, still require individuals to be actively involved to realise their objectives.

Organisations, however, have several significant characteristics that differentiate them from mere collections of individuals. They may have different time horizons in living longer than individuals who pass through the organisations which continue through time with a life of their own. Organisations may also have resources and power over and above those simply controlled by their individual members, partly because these organisations are recognised externally to have a separate legal identity. The resources of an organisation perhaps can only be mobilised through a consensus of its individuals, rather than by individual whim. so that the power of organisations is perhaps greater than the sum of its individuals' power.

Organisations may also develop their own values and perhaps therefore promote a subtle and unnoticed shift in the values of their members. Loyalty to the organisation, whether government agency or pressure group, may affect behaviour. Collective 'sticking together' may mean individuals pursuing what is not necessarily in their best interest, for the sake of collective progress or mutual self-defence. Ultimately, therefore, organisations develop distinctive behaviour characteristics reflecting their history and traditions, their accumulated experience and a collective perception of their rôles beyond those laid down formally, perhaps in law. They also form perceptions of their effectiveness, and a view as to the rôles of their members, all related to the underlying values of the dominant group within the organisation. Finally, they also develop mechanisms designed to sustain and perpetuate the organisation itself.

This distinctiveness of organisations, as opposed to collections of individuals, concerns both their internal operations and their externally oriented political behaviour, which routinely transcend the influence and power of individuals. Individuals are affected in turn by their organisations and by the external economic and political pressures they face.

All this means that in seeking comprehensive explanations for flood alleviation and land drainage policies or activities we should be wary of seeking these explanations in individuals' perceptions and actions. These individuals, whether farmers, floodplain dwellers, flood victims, land-drainage engineers or government officials, may well be pawns in a process that they do not recognise or understand correctly, and which they certainly cannot control. The institutional context dominating decisions and economic trends allows for individualism, but only to a limited and strictly controllable extent.

OTHER INSTITUTIONAL COMPONENTS: LAW, FINANCE AND STRUCTURES

The law is most significant in defining ownership and providing a collective memory of past agreements and decisions. Legislation also codifies the formal rôle of organisations and individuals. It also bestows power on those given legal rights and reflects the values of the society involved.

The law, however, is not a neutral or separable instrument of government but reflects the interests of those ruling the country concerned. It thus strives to uphold these interests, not least by tending to preserve the *status quo*, rather than seeking change by actively promoting alternative viewpoints. The law can change, of course, when those controlling the institutions involved see benefit in change. The evolving pattern of legislation and common-law rulings thus illuminates and encapsulates the changing influence and interests of those with dominant power. This analysis is not a criticism of the law as an institution but merely a recognition of its reality and purpose.

How finance is raised and spent may be prescribed by law. Both operations are crucial to policy formulation and decision making since money is the principal driving force behind construction programmes for flood alleviation and agricultural investment as in field drainage. Financial limits to revenue expenditure, and capital for private and public investment, depend ultimately on the state of the national or even the international economy. The priorities of government or others with political power determine the allocation of monies to different areas, including land drainage. Such allocation is based on political bargaining and a perception by those in power of national or local needs.

We thus need to evaluate the rôle and effects of the law in our particular environmental policy field, as well as the mechanisms and effects of financial and economic arrangements. The interlocking formal and informal structure of organisations also needs analysis. This is because the locus of an institution or organisation within the various hierarchies involved will affect its sphere of influence and power. Informal and formal relationships between organisations will in turn reflect this power, the exercise of which will require interorganisational co-ordination, delegation and bargaining. Within the organisational structure are people, some in government and some in other agencies, who can perhaps influence and sometimes change the legal and administrative rules by which the total system operates. Others have less dramatic but still important rôles such as co-ordinators, delegators, bargainers or publicisers. The example of our analysis is Britain, but the process of analysing these features is applicable elsewhere and in each country a different institutional context to land-drainage policies and decisions will reflect different cultural histories.

The legal framework

HYDROLOGICAL AND LEGAL DEFINITIONS

In hydrological terms 'drainage' embraces the land phase of the water cycle from the moment precipitation reaches the earth's surface to its return to the atmosphere through evaporation, evapotranspiration, or its disposal as surplus water to the sea (Ward 1975). In Britain the term 'land drainage' thus includes the full spectrum of physically interdependent drainage problems from agricultural soil drainage improvement to urban flood alleviation. The term, however, is traditionally most closely associated with the drainage of agricultural land. This is clearly reflected in official statements about the prime objective of 'drainage', such as (Water Space Amenity Commission 1980a):

> controlling and maintaining the water table in agricultural land to enable its maximum use for food production, and the disposal of surface water runoff and effluents without the creation of water flooding problems in agricultural land and urban areas.

Other official objectives of 'drainage' are stated as defence against sea water and arrangements for flood forecasting, warning and emergency action. The construction of gauges, sluices, spillways, by-pass channels and pumping stations, as well as communication with the public and the media, are the necessary related, but secondary, activities.

Land drainage excludes surface water sewerage and highway drainage by means of artificial pipes and culverts. It also excludes the protection of vulnerable parts of the coast from erosion as opposed to flooding (National Water Council 1978). Thus the official, legal and administrative definitions of 'land drainage' reflect the provisions of land drainage legislation rather than hydrological definitions or relationships. Statutory definitions do not emphasise the agricultural aspects as much as do the current administrative definitions and procedures.

In England and Wales the Land Drainage Act 1976 is the principal statute in the field of flood alleviation and land drainage (see Table 2.1). Under Section 116 of the Act 'drainage' includes: 'defence against water (including sea water), irrigation, other than spray irrigation, and warping'. Warping is the process of flooding low-lying land so that alluvium is deposited thereby adding to the land's fertility. Also: 'land drainage means the drainage of land and the provision of flood warning systems'. Section 17 of the Act expands this definition by stating that drainage works include the improvement and maintenance of existing watercourses and the construction of new ones. Section 81 of the Land Drainage Act 1930 excluded the distinction between irrigation and spray irrigation and made no mention of flood warning systems. However, both spray irrigation and flood warnings have become more important since 1930 and they are included within Section 38 of the Water Act 1973. This illustrates how the legal definition of land drainage has evolved as needs have changed.

PRINCIPLES OF BRITISH LAND DRAINAGE LAW

British land drainage law is a complex mixture of statutory and common law provisions supported by case law history. Statutes are passed by parliament to modify or consolidate existing legislation. Case law is essentially a series of precedents from judgements in legal cases when the statute and common law are applied and interpreted.

Traditionally, British law evolves gradually. It reflects the wisdom accumulated through experience and develops as needs and problems change. British land drainage law is no exception and has remained remarkably stable over time. Its long history dates back to the establishment of the Commissioners of Sewers in 1427 and the Bill of Sewers in 1531 (Wisdom 1975). The most important statute law to be enacted in the modern era was the Land Drainage Act 1930. This repealed provisions for the drainage districts in the Land Drainage Act 1861 and the general statutes relating to the Commissioners of Sewers. The current legal complexity has developed out of the long-standing need to tackle the many conflicts between the community's need for drainage, the restriction of individuals' rights necessary for this common good, and the obligations upon individuals not to hinder the drainage of their neighbouring areas.

British land drainage law thus is based upon a number of important but remarkably poorly understood principles. These fundamentally affect the influence, decisions and behaviour of both individuals and organisations. The principles are closely related to some of the fundamental aims of a free-enterprise orientated democratic society. As such they are controversial, reflecting the myriad interpretations of such aims when these are applied to particular issues or circumstances. The principles nevertheless underlie the organisational structure for flood alleviation and land drainage. In particular they define the powers of the agencies concerned and the nature of the financial arrangements.

The first of these principles is that responsibility for land drainage rests first and foremost with the individual riparian owner. In British common law all land belongs to someone, including land adjoining or under a watercourse. The landowner of the banks of a natural stream is the 'riparian owner' and has riparian rights (Wisdom 1975, pp. 83–9), which many individual landowners in Britain retain. The principle behind successive statutes has been to retain these rights where possible and, in consequence, to leave the ultimate responsibility for land drainage with property owners (Wilkins 1980) rather than with the state. The rights of riparian owners therefore remain ancient and powerful. They have the power to undertake drainage improvements on their land and the law limits, where necessary, the powers of drainage authorities to carry out these improvements. By retaining their rights, however, riparian owners have legal responsibilities. They are responsible for drainage on their land and they must not obstruct natural flow or cause an obstruction that will adversely affect a neighbour.

To help settle disputes a complex history of case law is associated with riparian rights. For example, a riparian owner has no right to build a 'mount' (i.e. a levée) which would in times of ordinary flood flow 'throw the river water on to the ground of an owner on the opposite bank so as to overflow and injure him' (Menzies v. Breadlbane 1828). On the other hand, in an extraordinary flood an owner may fence off his land in order to ward away the danger without regard for the consequences of his actions for his neighbour in turning the flood water away (Nield v. London & North Western Railway 1874; Lagon Navigation Co. v. Lambeg Bleaching Co. 1927) (Wisdom 1975).

Where it has been judged unreasonable to expect a riparian owner to be responsible for land drainage, or where problems of co-ordination are concerned, common law riparian rights have been altered and thereby reduced by statute law. Thus a riparian owner who fails to act in the interests of the common good may be compelled to do so. Powers have also been vested in state agencies. The Water Authorities in England and Wales therefore have the power to undertake land drainage improvements on major rivers or watercourses which flow to the coast, since it could not be expected that the riparian owners here should carry the burden of providing drainage for all the upstream areas.

To summarise, the retention where possible of individual riparian rights and obligations is fundamental to British land drainage law. These riparian rights remain powerful but the growing complexity of land drainage problems and needs has necessitated the strengthening of the powers of intervention and, in consequence, the erosion of common law riparian rights.

A second principle of land drainage law is that land drainage powers should be permissive rather than mandatory: their use is optional rather than obligatory. Having these permissive rather than mandatory land drainage powers for state agencies such as Water Authorities and local authorities is a recognition of the individual's rights and responsibilities; it also serves to limit the duties of the state. However, this situation can generate considerable public misunderstanding and frustration. The necessarily permissive powers bring the temptation, particularly during periods of financial restraint, for drainage authorities without statutory duties to hope that problems will be solved by others: to 'pass the buck'. This leads to difficulties and delay and is particularly troublesome in the case of urban flooding problems where District Councils have difficulty in financing their contribution to large schemes.

Making drainage authorities' powers mandatory, however, would be problematic for a number of reasons. For example, some flood alleviation schemes simply are not economically justifiable, and in any case complete flood protection is impossible. Furthermore, it would be impossible to force farmers to invest in drainage. It would therefore be most unwise to give the public power to require drainage authorities to 'solve' their flooding or drainage problems, much though this might appear sound in cases of delay or inaction. An alternative approach to this problem might be for more co-operation amongst authorities (Wilkins 1980). In the meantime care needs to be taken when talking of the 'responsibilities' of a particular agency. In law this agency may not necessarily have obligations and it cannot be correctly accused of not fulfilling its statutory responsibilities if it chooses not to exercise its permissive powers (Wakelin 1980).

A third principle of land drainage law and procedures is that land drainage is predominantly a local problem and decisions about it should be made locally. Because individual landowners are legally responsible for draining their land, and also because each flooding or agricultural drainage problem is seen as unique, land drainage is viewed as a local rather than a regional or national issue. Exceptional floods periodically cause disasters which invoke national concern and central government response, such as in 1947, 1953, 1968 and 1982, but comparatively minor individual flooding and drainage problems are numerically far more important and typical. Thus where local interests are affected the official view is that decisions should be made by those most concerned. The result is that land drainage decision making is locally based, within catchment areas, for both urban flooding and agricultural drainage.

The fourth principle implicit in much legislation and its application is that those who benefit from land drainage or create a need for it should pay accordingly. This principle is embodied in Section 30 of the Water Act 1973 and in other principal legislation on land drainage (Table 2.1). Where improvements are made to private property, including improved land drainage or its protection from flooding, the law holds that it is not practicable or equitable for Water Authorities to shoulder the financial burden. This principle extends to other water services, such as the provision of private sewers or plumbing within the boundaries of private property (Wilkins & Lucas 1980), and dictates that community investment should not result in undue private gain.

Table 2.1 Principal legislation affecting flood alleviation and agricultural land drainage in Britain.*

England and Wales	Land Drainage Act 1930 Agriculture Act 1937, Section 15 Drainage Rates Act 1958 Land Drainage Act 1961 Drainage Rates Act 1962 Drainage Rates Act 1963 Agriculture (Misc. Provisions) Act 1968 Agriculture Act 1970, Part V Water Act 1973, Sections 9, 19, Schedule 5 Land Drainage Act 1976 Land Drainage (Amendment) Act 1976 Wildlife and Countryside Act 1981
Scotland	Land Drainage (Scotland) Act 1930 Land Drainage (Scotland) Act 1935 Agriculture Act 1937 Land Drainage (Scotland) Act 1941 Land Drainage (Scotland) Act 1958 Flood Prevention (Scotland) Act 1961 Agriculture Act 1970

*Successive statutes repeal either the whole or sections of previous Acts. For repeal details see the individual law statutes. Excluded above are the Miscellaneous War Provisions Acts.

The organisational structure in England and Wales

FIFTY-FIVE YEARS ON

The Land Drainage Act 1930 'put land drainage (including flood alleviation) in England and Wales on its feet' (Cole 1976). The Act established drainage districts, drainage boards and central government grants for land drainage investment. These drainage boards included Catchment Boards responsible for the single drainage function. Drainage boards or Internal Drainage Boards were created for areas outside the area of the Catchment Boards, to embrace such areas 'as derive benefit or avoid damage as a result of drainage operations'. In this way a Ministry of Agriculture and Fisheries letter sent in 1933 to the River Medway Catchment Board defined the 'Medway letter line' as the area of Internal Drainage Boards and, incidentally, the benefit area for agricultural enhancement calculations (pp. 110–15).

The River Boards Act 1948 brought changes to the 1930 Act. Catchment Boards were superseded by 32 River Boards together with the pre-existing Thames and Lea Conservancies. Drainage districts were included in River Board areas and became Internal Drainage Districts under the jurisdiction of Internal Drainage Boards. River Boards were also given powers over fisheries and pollution control. The Land Drainage Act 1961 did not affect the functions of the drainage boards although the Water Resources Act 1963 finalised the repeal of the 1948 Act by replacing the River Boards with 27 River Authorities which were given new functions relating to water resources.

The Water Act 1973 replaced the River Authorities with ten regional Water Authorities (Fig. 2.1). However, the Water Authority Land Drainage Committees are now the direct descendants of the Catchment Boards (Fig. 2.1) both as to the areas covered and their mode of operation. Government policy in 1977 indicated that changes in the organisational structure for land drainage were not then contemplated (Department of the Environment *et al.* 1977) but during 1982–3 an interdepartmental review began to scrutinise fundamentally all aspects of land-drainage administration.

Despite the frequent reorganisations in the water industry, the principal features of the organisation structure for land drainage have remained comparatively stable for over 50 years. This is due to the general acceptance of the legal principles discussed above, which dictate important elements of this structure including the local or regional control within Catchment Boards, Water Authorities or Internal Drainage Boards. In addition the structure has been strenuously defended by the farming and landowning land drainage interests who believe it to be appropriate and efficient. Pressure to alter the structure has arisen in the past but these interests have lobbied successfully for its retention.

For example, in negotiations leading to the Water Act 1973 a vigorous campaign was mounted to keep land drainage separate from the new multifunctional Water Authorities and to prevent the takeover of land drainage responsibilities by the Ministry of Housing and Local Government (Richardson *et al.* 1978). Those involved included a coalition of the National Farmers' Union, the Country Landowners' Association, the Association of River Authorities, the Association of Drainage Authorities and senior civil servants in the Ministry of Agriculture, Fisheries and Food. This powerful agricultural alliance was termed the 'MAFFia' by their opponents and the River Authorities most concerned to retain separate land drainage functions worked together as the 'wet seven'. In the event a compromise was reached. Despite representing a structural anomaly the Ministry of Housing and Local Government conceded a continued separate organisation for land drainage within the Water Authorities, probably because of the lack of parliamentary time. The overall supervisory rôle of the Ministry of Agriculture, Fisheries and Food therefore remains today and statutory Land Drainage Committees within Water Authorities carry on where the Catchment Boards began. The all-important system of grant aid for land drainage was allowed to continue and the Ministry nominates members to the Water Authority boards to safeguard agricultural interests.

THE MINISTRY OF AGRICULTURE, FISHERIES AND FOOD AND THE DEPARTMENT OF THE ENVIRONMENT

Under Section 1 of the Water Act 1973 a duty of the Minister of Agriculture, Fisheries and Food is:

> to secure the effective execution of so much of that policy (i.e. the

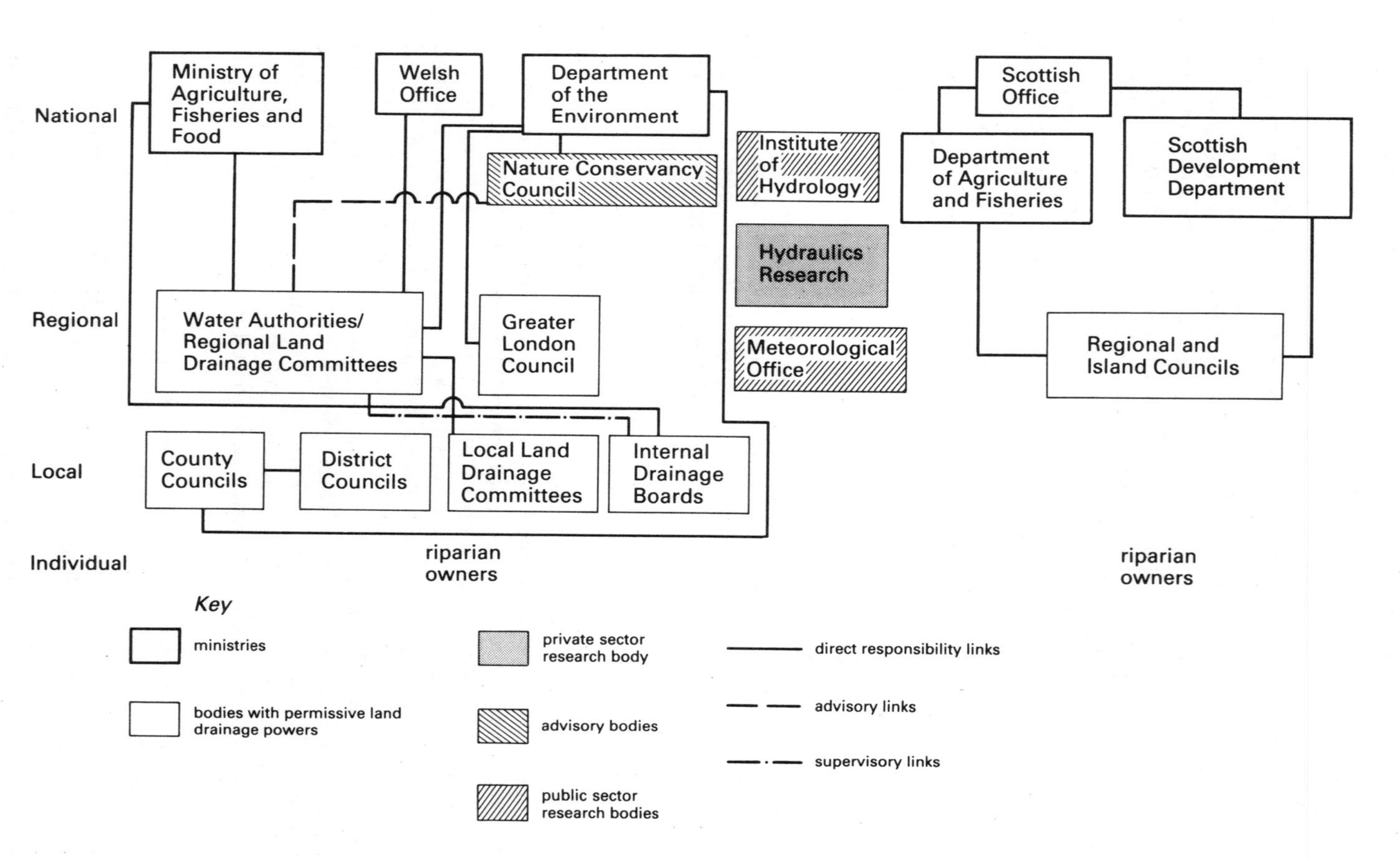

Figure 2.1 A simplified organisational structure for flood alleviation and land drainage in Britain.

national policy for water) as relates to land drainage and to fisheries in inland and coastal waters.

Prime responsibility for flood alleviation and land drainage thus rests with the Minister of Agriculture, Fisheries and Food, although some matters are the joint responsibility of the Minister and the Secretary of State for the Environment. These particularly concern navigation and the planning of new urban development in flood-prone areas. For example, a set of administrative rules issued originally by the then Ministry of Housing and Local Government (now the Department of the Environment) is of particular importance to land drainage. The latest circular 17/82 concerns the possible adverse effects of development in flood risk areas. In common with previous circulars, issued in 1962 and 1969 respectively, the advice to local town and country planning authorities is to consult Water Authorities about development in floodplains and about the effect of runoff from developments that may exacerbate flooding downstream.

The Ministry of Agriculture, Fisheries and Food (and in Wales, the Welsh Office) exercises overall co-ordination of flood alleviation and land drainage, as part of their wider responsibility for agricultural policy and development. They have the powers to give direction on land drainage matters and also to give grants to drainage authorities towards the costs of their schemes. Ministry staff also co-ordinate opinion concerning land drainage, both formally and informally, for example through promoting the annual Cranfield Conference of River Engineers at which research results are presented and grievances about government policy are aired.

The Ministry includes two divisions – one concerned with land drainage and the other with fisheries – and also a Land Drainage Service as part of the Agricultural Development and Advisory Service. The latter advises farmers and Land Drainage Committees on the potential for field drainage and main river works and upon the system of grants available for drainage work (Agricultural Development Advisory Service 1974–7). All major land drainage programmes and plans must be submitted to the Ministry which assesses both the engineering design and the economic analysis undertaken by the Water Authority, local authority or others to justify both urban flood alleviation and agricultural land drainage schemes (Ch. 4). In this way central control over financial aspects of local decision making is carefully retained, but ultimate responsibility for policy and implementation remains usefully removed from central government. In theory the Ministry's assessment only concerns whether it will give grant aid to the land drainage scheme and does not ultimately determine whether the scheme will go ahead. However, few schemes would proceed without grant aid and the Ministry approval it conveys.

As is traditional in the British civil service, the Ministry takes a relatively conservative approach to land drainage affairs. The Ministry is wary of appearing to be critical of the autonomous Water Authorities and of joining, even within the water industry, an open and public debate of land drainage policy. Rather than adopting a positive leadership rôle towards policy review – which might be one method of strengthening its land drainage rôle – the Ministry appears to adopt a more defensive behaviour. Thus a debate is generally only joined through necessity, when it has already been initiated within the Water Authorities or has been raised to a politically significant level by some other external body.

This traditionally cautious British approach in which changes are only made somewhat reluctantly and at the margins of the *status quo* – if at all – contrasts markedly with the more aggressive leadership rôle taken, for example, by the US Army Corps of Engineers in self-critically debating its rôle in promoting non-structural flood mitigation measures. In Britain policy change, and even policy review, tends to be resisted within government agencies or is at best made only following careful 'closed door' discussions with interested parties.

THE WATER AUTHORITIES' RÔLES

Under Section 1(1) of the Land Drainage Act 1976 Water Authorities 'shall exercise a general supervision over all matters relating to land drainage in their area.' This general supervision extends to all watercourses, both 'main' and 'non-main'. Under Section 9 of the Land Drainage Act 1976, 'main rivers' are those shown on the statutory main river maps held by the Water Authorities and the Ministry. The distinction is most important. Powers for undertaking drainage-related work on main rivers are exercisable by Water Authorities and by others with the Water Authority's consent. Water Authorities, however, have no such authority for other watercourses, where Internal Drainage Boards, District Councils or riparian owners have the necessary powers. The overall Water Authority supervision is exercised in part by enforcing bye-laws, by consenting to works on or in watercourses (Ch. 6) and by liaison with planning authorities responsible for development control. Such Water Authority powers are largely permissive, despite the statutory supervisory duty outlined above. Therefore, for example, under Section 32 of the Land Drainage Act 1976 Water Authorities have permissive powers but no duty to provide flood warning systems.

Under Section 34 of the Land Drainage Act 1976 drainage authorities can 'make such bye-laws as they consider necessary for securing the efficient working of the drainage system in their area.' These bye-laws concern the prevention of damage to banks, the opening of sluices and flood gates, the prevention of obstruction such as rubbish, and require those responsible to cut and remove vegetation in or on the bank of a watercourse. To maintain standards for watercourses where alterations could adversely affect the main river system downstream, Land Drainage Consents are required from the Water Authority to carry out the works. These Consents ensure that land drainage operations will not be detrimentally affected and flooding thereby caused.

The Water Authorities execute the majority of flood-alleviation and arterial land drainage schemes in England and Wales, usually through their rivers divisions. This involves hydrological and hydraulic modelling, engineering design, cost–benefit appraisal and the supervision of scheme installation and maintenance. To perform these tasks the Authorities maintain professional hydrologists and engineering staff who, as in most organisations, seek to protect their own employment or enhancement by ensuring a steady flow of drainage work. This 'survival' objective may take the form of careful and wise planning ahead or a more random 'search' for work in the form of viable schemes. It undoubtedly forms an important institutional factor in determining decisions to pursue a particular policy or to promote a particular scheme.

Water Authorities also have a longer term objective in seeking to control floodplain developments through consultation with town and country planning authorities. Following circular 17/82 and its predecessors, Water Authorities advise planning authorities on the wisdom of such developments (Penning-Rowsell & Parker 1974). In cases likely to increase the risk of flooding the developers should be told how to remedy the problem, perhaps by raising ground levels above known flood heights or by providing on-site flood storage. If the developer fails to agree satisfactory arrangements to tackle the problem the planning authority – but not the Water Authority – can refuse the relevant planning application or impose the necessary conditions (Penning-Rowsell 1981a).

Non-main river works may be carried out by the developer at his own expense by agreement with riparian owners. Where agreement is not possible the developer may ask the local authority to carry out the works and reimburse the authority accordingly. Main river works will normally be undertaken by the Water Authority, perhaps with a contribution from the developer. Currently the developer's contribution to watercourse improvements away from the developer's site is negotiable and in practice few worthwhile contributions can be agreed. Proposals have been made to make developers' contributions mandatory so as to compel developers who cause drainage and flooding problems to pay accordingly, as in line with the fourth principle discussed above (page 29). However, no legislation concerning this compulsory payment exists at present.

THE REGIONAL AND LOCAL LAND DRAINAGE COMMITTEES: THE POWER BASE

Water Authorities are statutorily required to discharge their land drainage functions via separately constituted and financed regional Land Drainage Committees. This means that the regional Land Drainage Committees, and not the Water Authorities, are the principal flood alleviation and land drainage decision-making organisations. Only where land drainage functions appear materially to affect the Water Authority's other water management rôles may the Authority give directions to their Regional Committees. Nevertheless in practice it is Water Authority staff who plan and undertake land drainage work, supervised by the Committees but also accountable to the Authority directors.

Under Section 4 of the Land Drainage Act 1976, and in line with the third principle underlying land drainage law, each regional committee delegates its powers to a number of local Land Drainage Committees. For example, the Anglian Water Authority's regional committee has established five local committees and Severn-Trent has a local committee for each of the Severn and Trent catchments. Rules for the constitution and membership of regional and local Land Drainage Committees provide for the representa-

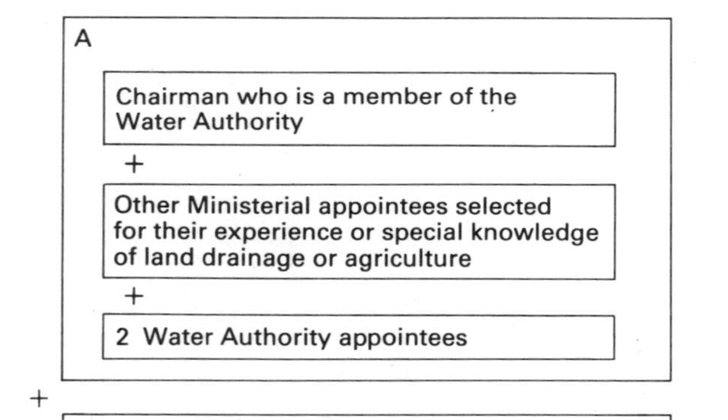

*Constituent councils may not be represented if, in the Minister's view, the land drainage revenue from that area is below an appropriate level

Figure 2.2 Statutory rules for membership and representation for Land Drainage Committees in England and Wales under the Land Drainage Act 1976 (Sections 2–5). The Thames Water Authority has slightly different rules because of the Greater London Council and the London Excluded Area.

tion of local interests (Fig. 2.2). However, only the County Councils and not the District Councils are considered to be constituent organisations because land drainage precepts (charges) are taken from the former and not the latter. Committee members are appointed rather than elected and, in practice, the rules make likely the domination of the committees by agricultural and landowning interests (Fig. 2.2). Although not excluded by the rules, the fishery and conservation interests often affected by land drainage are not automatically included (Penning-Rowsell 1980). In practice, fisheries' interests are commonly represented but conservation interests may not be.

THE RÔLE OF THE LOCAL AUTHORITIES

Under Sections 97–104 of the Land Drainage Act 1976 District and County Councils have powers for maintaining and improving non-main rivers. These powers are complementary to the Water Authorities' main river powers. The District Councils' powers are for preventing flooding or alleviating flood damage (Section 98) (Garner 1980). In practice District Councils are concerned with flood alleviation in both rural or urban environments, whereas County Councils are mainly concerned with agricultural land improvement and, under the Highways Act 1959, with road drainage to prevent highway flooding. County Councils have powers to undertake land drainage schemes at the request of owners and occupiers who will benefit from the scheme (Section 99). Following the fourth principle underlying land drainage law the cost of those schemes is apportioned amongst the beneficiaries. County Councils can also carry out land drainage works compulsorily for the improvement of agricultural land (Section 100) and they may also execute powers by agreement with, or by default of, a District Council (Section 98) (Wood 1980).

Following the second principle underlying land drainage law, local authorities' powers are permissive. Because of this, and the fragmentation of land drainage responsibilities amongst a large number of drainage authorities, the question of who should accept responsibility for tackling a particular flooding problem is often a vexed one. The meaning of permissive powers is often misunderstood leading to accusations of negligence in cases where none of the 'responsible' authorities appears to be shouldering the burden (Hitchenor 1980).

In addition to their drainage powers District Councils also have statutory development control powers. These powers are conferred under the Town and Country Planning Acts 1947, 1971 and the Town and Country Planning General Development Order 1977 (Cullingworth 1972). They are important where attempts to control floodplain development or increased urban runoff are concerned. The procedures that accompany these powers are laid out in the circular 17/82 discussed above.

A weakness of local authority involvement in flood alleviation is that their staff do not routinely encounter flood alleviation design problems as do their Water Authority counterparts. A likely consequence of this necessary devolution to a local authority level of the power to deal with local problems is thus a lower standard of professional competence which may result in inadequate 'solutions' or inordinate delays.

THE INTERNAL DRAINAGE BOARDS

Internal Drainage Boards are virtually autonomous co-operatives of farmers and landowners who have a strong interest in ensuring that drainage work continues (Hall 1978). Boards were first constituted in the 19th century under individual Acts of Parliament but gradually they have amalgamated and are now constituted under Sections 6 and 7 of the Land Drainage Act 1976. Internal Drainage Boards' powers are permissive and they 'shall exercise a general supervision of all matters relating to the drainage of land within their districts' which 'shall be such areas as will derive benefit or avoid danger as a result of drainage operations'. These areas are generally lowlands with special drainage problems and where collective benefit will be derived by a number of landowners from co-operation over drainage work (Fig. 2.3.)

There are almost 300 Boards in England and Wales covering 1.2 million hectares and district boundaries are determined by reference to the 'Medway letter'. The area of a Board is thus defined somewhat arbitrarily as being that agricultural land up to 8 ft (2.4 m) above the highest known flood level (5 ft (1.5 m) above ordinary spring tide in tidal areas) or, in the case of urban land, up to flood level. Under Section 1 of the Land Drainage Act 1976, Internal Drainage Boards are empowered to carry out work on all non-main rivers within their area. In practice, most Boards designate certain watercourses in their district for their regular maintenance and other minor watercourses are left to riparian owners to maintain or improve.

Rules concerning the membership and operation of Internal Drainage Boards are set out in Section 7 and Schedule 2 of the Land Drainage Act 1976. The members of the Boards are elected from those owning at least 10 acres (4.05 ha) of land in the Internal Drainage District, the occupiers of at least 20 acres (8.1 ha) of such land, the owners or occupiers of land of an annual value of £30 or more, or their nominees. Voting power is biased in favour of the large landowner: according to the assessable value of the property concerned an elector is entitled to between one and ten votes. Terms of office are normally 3 years and Boards must send an annual report to the Minister and each Water Authority, County or London Borough covered by part of the District.

The organisational structure in Scotland

The organisation of Scottish land drainage and flood alleviation is different from that in England and Wales and, in particular, the rights and responsibilities for drainage remain even more with the riparian owner. Nevertheless the state exercises both a financial and supervisory rôle through local authorities and the Scottish Office.

THE SCOTTISH OFFICE

The Scottish Office, and the Secretary of State for Scotland, is responsible for virtually all the duties and functions which in England and Wales come under the Department of the Environment and the Ministry of Agriculture,

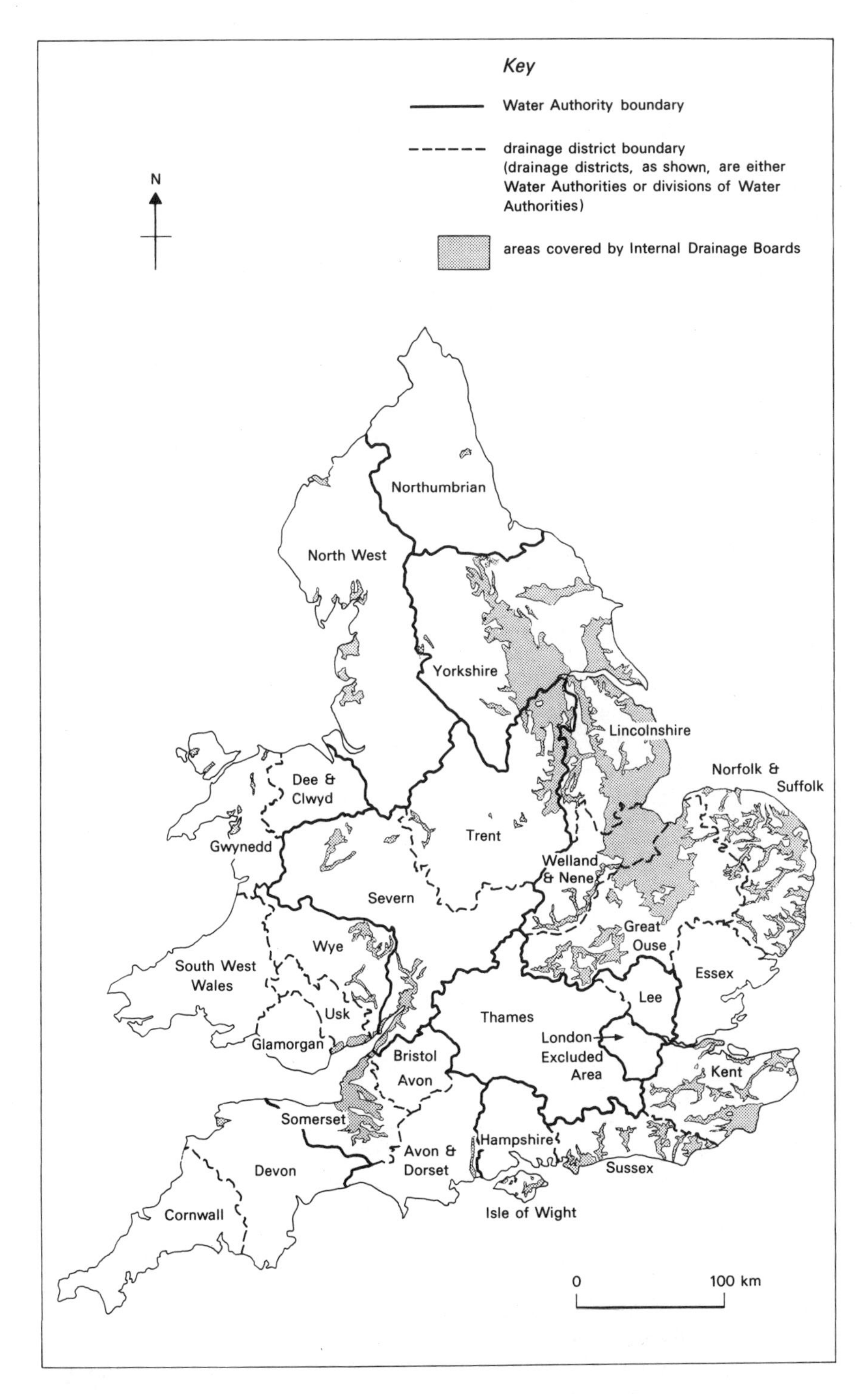

Figure 2.3 The boundaries of the Water Authorities, Local Land Drainage Committee areas, and the total extent of Internal Drainage Districts.

Fisheries and Food. Within the Scottish Office there are separate departments: the Department of Agriculture and Fisheries and the Scottish Development Department; both have functions in the drainage field. Agricultural land drainage schemes are the responsibility of the Department of Agriculture and Fisheries, whereas the Scottish Development Department oversees the urban flood protection work undertaken by the 12 Regional and Island Councils, which in this respect have rôles almost equivalent to those of the Water Authorities in England and Wales.

THE REGIONAL AND ISLAND COUNCILS

Flood prevention became the responsibility of the nine Regional and the three Island Councils in the 1975 reorganisation of Scottish local government and water services following the Local Government (Scotland) Act 1974. Previously, flood prevention was undertaken by many small burghs and councils. The Regional and Island Councils are multipurpose planning authorities and have responsibilities ranging from housing and education to water, sewerage and flood prevention (Parker & Penning-Rowsell 1980).

Urban flooding is apparently less of a problem in Scotland than in England and Wales. As a consequence of this lesser importance, flood prevention may come under a water and sewerage department of the Councils, as in the case of the Highland Regional Council. The Flood Prevention (Scotland) Act 1961 relates only to non-agricultural land and under this Act Regional and Island Councils are responsible for identifying, evaluating and designing schemes for grant aid by the Scottish Development Department. In 1977 and 1978 the number of schemes submitted under the 1961 Act was zero and four respectively and the total approved expenditure for the four schemes was only approximately £0.3 million for which a grant of 30 per cent was generally paid. The Scottish Development Department has the power to confirm or reject schemes, which are not restricted to inland areas but can include estuaries, but not coastal areas, covered by the Coast Protection Act 1949.

THE RIPARIAN OWNER IN SCOTLAND

In contrast to England and Wales Scottish riparian owners are responsible for proposing agricultural drainage schemes and the necessary arterial drainage improvements. Under the Land Drainage (Scotland) Act 1958 an owner or group of owners may apply to the Secretary of State for an 'Improvement Order' for an area suffering from flooding, inadequate field drainage or land erosion by rivers. The Department of Agriculture and Fisheries then issues an order, if the improvement is judged to be warranted, to specify the area to be improved and the works necessary and their cost. Where multiple landownership is concerned, the Improvement Order can set up a Committee to oversee the landowners' responsibilities, particularly with regard to maintenance.

The Act provides for grant aid for the capital costs involved – up to 50 per cent but generally increased to 60 per cent under the Farm Capital Grant Scheme – and requires the owner to maintain the completed scheme in good condition. The majority of schemes carried out under this legislation are large-scale arterial drainage projects undertaken jointly by several landowners, although the annual average total grant is only some £50 000. Both costing and design are carried out at central government level, by the Department of Agriculture and Fisheries, as the individual landowner cannot be expected to have these skills.

Non-statutory organisations and the wider political environment

The institutional context of flood alleviation and land drainage extends beyond statutorily defined government and specialist agencies. A number of other organisations can be identified which together comprise components of the wider political environment of land drainage. These organisations are significant in policy analysis because their expertise influences decisions and because they have important interests in promoting flood alleviation and perpetuating or reducing the pace of agricultural drainage.

RESEARCH ORGANISATIONS

All organisations with relevant research expertise tend towards favouring the continuation of flood alleviation and land drainage, although they may not directly pursue these ends. These organisations undertake research with contracts let by the government-financed Research Councils, by the Ministry of Agriculture, Fisheries and Food or the Water Authorities, or from others that broadly defend land drainage interests. The research itself may therefore be oriented towards strengthening the positions of contracting bodies or the perpetuation of their drainage policies and programmes.

In addition to the Universities and Polytechnics, and the Agricultural Development Advisory Service of the Ministry, four research organisations in Britain play significant rôles in the field. The Institute of Hydrology completed during the 1970s a major study on floods (Natural Environment Research Council 1975) and this has led to improved hydrological data and methods for estimating flood frequency and magnitude (Ch. 3). More recently the Institute's research has included improvements of the basic statistical procedures contained in the original *Flood Studies Report* in order to improve the design of flood protection works and flood warning systems. The Institute is also involved in a range of research projects including some on the effects of catchment urbanisation on runoff and the hydrological effects of field drainage.

The Institute of Oceanographic Studies is the main source of expertise in predicting coastal flood magnitude and frequencies. Recent research has included the modelling of tidal surges, extreme sea levels and wave patterns. In this way the Institute is contributing towards improvements in the forecasts issued by the Storm Tide Warning Services. The Hydraulics Research Station was until 1982 part of the Department of the Environment but has since become a private sector research agency, now called Hydraulics Research. This agency undertakes both mathematical and physical modelling of flood-prone areas to determine parameters such as the increased channel capacity necessary for flood-alleviation schemes. Recently research has investigated methods of protecting sea defences, techniques for optimising urban storm drainage design, and improved methods for measuring river flow velocities. The Meteorological Office provides a weather forecasting service which is important in estimating the likelihood of floods and in issuing accurate flood warnings.

PROFESSIONAL INSTITUTIONS, ASSOCIATIONS AND INTEREST GROUPS

The land drainage field involves a wide range of other organisations. A pluralistic model of society sees policy making as the interaction of a number of bodies acting as pressure groups seeking to promote their viewpoint and to develop their power to influence decisions in their interests. To evaluate this model's usefulness we need to look critically at the real power of such groups.

The Institution of Civil Engineers and the Institution of Municipal Engineers both seek to maintain professional standards and to provide a relatively closed forum for debating issues of mutual importance to land drainage and flood alleviation. As well as their educational rôle these bodies collate professional viewpoints upon land drainage matters, amongst others such as the need for institutional reform or research. They seek by lobbying to bring about the change or to maintain the *status quo*. As professional institutions they naturally favour a flourishing engineering profession and therefore they may be expected on balance to favour a steady flow of land drainage work. They will also seek the continued domination of the engineer in this field.

Land drainage is also strongly supported by the National Farmers' Union and the Country Landowners' Association which are particularly powerful and influential; the Scottish Federation of Landowners is a Scottish counterpart. This influence is strengthened by the strong community network of interests aimed at promoting agricultural drainage so as to maintain and increase the productivity and profitability of private farming assisted by government grants. Such a network of interests is made formidable by the overlapping membership of these interest groups, Land Drainage Committees and the Internal Drainage Boards thus making the farming lobby a powerful alliance.

The Country Landowners' Association and the National Farmers' Union thus mobilise their influence through their representation on important regional or national advisory committees and professional bodies. Both are well represented on Land Drainage Committees and to a lesser extent on the Water Authorities themselves. The members of Water Authorities appointed by the Minister of Agriculture are selected from nominations made by the Country Landowners' Association and the National Farmers' Union amongst other bodies. Both interest groups are also very active in lobbying the government and the European Community organisations concerned with setting agricultural prices and determining export subsidies.

Because it is so strongly to their advantage the Country Landowners' Association and the National Farmers' Union both maintain a keen interest in retaining the existing institutional arrangements for land drainage. This includes the present systems of Land Drainage Committees, local administration and financing with grant aid. Both also support the continued overall supervision of land drainage by the Ministry of Agriculture and the Minister's involvement in appointing members to Water Authorities and to Land Drainage Committees. It is clear that the Ministry, the Country Landowners' Association and the National Farmers' Union often have considerable coincidence of interests. This was exemplified during the consultations over the Water Bill in 1972, with the pressure groups defending the *status quo* and the Ministry defending in parallel its own administrative territory (Richardson *et al.* 1978). A joint defence or response – supported by the Ministry – to the growing public disquiet over the environmental impact of farming has emphasised their caring for the countryside (National Farmers' Union and Country Landowners' Association 1977).

So far, the main challenge to the land drainage *status quo* has come from the numerous environmental pressure groups collectively representing an environmental 'movement'. Undoubtedly the rise of this movement reflects a substantial shift in public values in the 1960s and 1970s and, of course, it is concerned with much more than just land drainage. Nevertheless during the late 1970s and 1980s agricultural drainage schemes and policies have become a major focus of controversy and bitter disagreements (Chs 5 & 6). The main groups active in opposing current practices have been the Council for the Protection of Rural England, with its emphasis on landscape conservation, the Royal Society for the Protection of Birds, and the government conservation agency the Nature Conservancy Council. Individual local amenity societies or branches of national organisations have also sprung up to oppose particular agricultural drainage schemes and, to a lesser extent, certain urban flood alleviation and sea defence works.

Such opposition has undoubtedly had an impact upon official opinion and policy, as particularly encapsulated in the Wildlife and Countryside Act 1981. In this respect the pressure-group politics surrounding a particular land-drainage scheme need to be understood for a complete policy analysis. However, it remains difficult to determine accurately the real power of such environmental groups *vis-à-vis* the undoubtedly powerful agricultural organisations. This is largely because it is difficult methodologically to separate out the contribution of environmental groups to the evolution of events. All it is often possible to conclude is that they *appear* to be influential in particular situations, such as at public inquiries or in drafting the Wildlife and Countryside Act, but this appearance could easily under- or over-value their real effect.

Further legal obligations and institutional features

CONSERVATION AND FISHERIES

The law recognises that land drainage and flood alleviation works have important consequences for nature conservation values (Ch. 5). However, in the conservation legislation the obligation of responsible organisations is often somewhat vague, specifying their obligations only 'to have regard to' environmental matters and this is not as effective as the protection of a legal right in law. Thus, for example, under Section 48 (1) of the Wildlife and Countryside Act 1981, in considering any proposals relating to the discharge of their functions, Water Authorities:

(a) shall, so far as may be consistent with the purposes of this Act and of the Land Drainage Act 1976, so exercise their functions with respect to the proposals as to further the conservation and enhancement of natural beauty and the conservation of flora, fauna and geological or physiographical features of special interest;
(b) shall have regard to the desirability of protecting buildings or other objects of archaeological, architectural or historic interest; and
(c) shall take into account any effect which the proposals would have on the beauty of, or amenity in, any rural or urban area or on any such flora, fauna, features, buildings or objects.

There are also statutory obligations under the Countryside Act 1968 (Section 11) and the Countryside (Scotland) Act 1967 ensuring that in exercising their functions under any Act every Minister, government department and public body, including Water Authorities and Regional Councils, 'shall have regard' to the desirability of conserving the natural beauty and amenity of the countryside. Further legal obligations exist for the Nature Conservancy Council to notify Water Authorities of the existence of Sites of Special Scientific Interest (SSSI) and for Water Authorities to notify conservation bodies of their drainage proposals (Ch. 6).

Game fishing and local anglers' associations, often affiliated to the National Anglers' Council, are relatively powerful pressure groups within the water industry. This power partly derives from the large numbers participating in the sport and the groups are sensitive to any adverse effects on fish stocks by flood alleviation or drainage works (Davies & Parker 1982).

Thus under Section 113 of the Land Drainage Act 1976 drainage authorities and local authorities must 'have due regard' for the interests of fisheries, including sea fisheries, when they carry out land drainage works. There are also important provisions in the Salmon and Freshwater Fisheries Act 1975. For example, if in carrying out drainage works salmon and trout are hindered, scared or prevented from passing through a fish pass, then an offence has been committed. Where estuarine and sea fisheries are concerned land drainage or sea defence works are also subject to the protective provisions of Sea Fisheries (Shell Fish) Act 1967 and the Dumping at Sea Act 1974.

SEA DEFENCE

Protecting the coastline from erosion is not legally part of land drainage. Nevertheless, where the coastline is not backed by high land, sea defences may be necessary to prevent both erosion and flooding. Both matters are governed by the Coast Protection Act 1949 which extends also to Scotland.

The coastline is split into stretches each with a coastal protection and sea defence authority which include maritime District Councils and the Water Authorities. Difficulties of co-ordinating sea defence works are sometimes encountered where more than one authority is involved and the principal objective of a scheme – whether erosion or flood prevention – affects the rate and source of government grant aid. Large-scale sea defence works are particularly important on Britain's east coast where a substantial programme of flood alleviation defences has reduced the chance of repetition of the disastrous 1953 floods.

NAVIGATION

Provision for navigation sometimes has important consequences for flood alleviation and land drainage, and vice versa. The responsibility for maintaining rivers, tidal creeks and ports in a navigable condition rests with a variety of public bodies including Water Authorities and Port Authorities. For example, in the Norfolk Broads the navigation authority functions are performed by the Great Yarmouth Port and Haven Commissioners. These Commissioners possess permissive powers to deepen, dredge, cleanse and scour the Haven and the rivers. They also have powers to preserve them from encroachment, to preserve public rights of way, to improve navigation facilities such as landing stages and to promote bye-laws for controlling boating. Insofar as navigation authorities like the Commissioners seek to maintain and improve channels, for example through dredging or weed cutting, they can at the same time assist flood alleviation and affect agricultural drainage. However, not all flood protection and land drainage works are necessarily of assistance to navigation: the Thames navigation requirements posed considerable constraints on both the siting and design of the Thames tidal surge barrier (Ward 1978, Horner 1978, 1979).

LONDON

The London Excluded Area covers about 1040 km^2 and includes most of Greater London and parts of adjacent administrative areas. Because metropolitan administration and land drainage problems have been considered unique the London Excluded Area and the Greater London Council have been subject to separate legal provisions. Nearly all of the Area is within the catchment of the Thames Water Authority, but this Authority is 'excluded' from exercising its function here. The situation is anomalous, resulting from a compromise in the bargaining leading to the Water Act 1973, and may well not survive future reorganisations of London's local government.

Within the Area drainage functions are exercised by the Greater London Council, London Boroughs and the Common Council of the City of London. 'Metropolitan watercourses' and 'main metropolitan watercourses' are defined in Schedule 5 of the Land Drainage Act 1976 and in The London Excluded Area (Designation of Watercourses) Order 1976. On the 'main metropolitan watercourses' drainage functions are exclusively exercised by the Greater London Council. On other 'non-main metropolitan watercourses' both the London Boroughs and the Greater London Council may carry out these functions.

The Greater London Council's land drainage powers are similar to those of Water Authorities in being permissive, but the Council also has mandatory planning powers to control development affecting watercourses named in local enactments, such as the River Ravensbourne, etc. (Improvement and Flood Prevention) Act 1961, and the London and Surrey (river Wandle and river Graveney) (Jurisdiction) Act 1960. Separate legislation also exists for tidal watercourses in the London Excluded Area. Under The Thames River (Prevention of Floods) Acts 1879 to 1962 the Greater London Council has powers of control over developments affecting flood defences. Extensive and very important powers concerning the construction and operation of the Thames Barrier were separately provided in the Thames Barrier and Flood Prevention Act 1972.

The National and European economic context

National economic policy and fortunes are of crucial importance to both government agencies and private entrepreneurs engaged in flood alleviation and land drainage. Britain's declining economic performance during the 1970s set the scene for International Monetary Fund intervention in 1976 and thereafter the introduction of strong monetarist and related policies. Public expenditure came under increasing pressure during both Labour and Conservative administrations. Agencies proposing flood alleviation schemes have consequently come under the tighter financial control of central government, facilitated in England and Wales by the reorganisations under the Water Act 1973 (Penning-Rowsell & Parker 1983). Capital investment ceilings have declined and expenditure proposals have received progressively closer scrutiny by both central government departments including the Treasury and by local ratepayers' representatives seeking to control rate rises at times of static or falling living standards.

Within this period, at least until 1983, the flood alleviation budget of the Ministry of Agriculture, Fisheries and Food has escaped largely unscathed owing to the political support from the agricultural drainage interests. Other Water Authority investment and local authority expenditure, in contrast, has been cut back substantially thus affecting storm sewerage investment, non-main river flood alleviation and sea defence schemes. However, increased control over public expenditure has not only resulted in expenditure cutbacks. In November 1982 the Conservative government instructed local authorities to spend more on capital programmes in order, indirectly, to counter the seriously depressed economic and employment situation. Nevertheless, decisions to pursue particular flood alleviation or sea defence schemes may not be directly traceable to these fluctuations in expenditure policy. The system of tighter central scrutiny and control remains, however, as does the encouragement of a more free-enterprise ethic where it is hoped this can result in economic regeneration. No systematic evidence exists, but this in turn might be important in understanding why certain urban development on floodplains has been allowed to occur to create employment where job opportunities might otherwise have gone elsewhere.

However, probably the most important area in which the wider economic environment is crucial to an understanding of policies and decisions concerns agricultural land drainage. Although ultimately the decisions are those of the individual farmer or entrepreneur the driving force is the economic climate created for farming by government support of agriculture of which the current form comprises the Common Agricultural Policy (CAP) of the European Community.

To put this in context, agriculture in Britain contributes some 3 per cent of gross national product and employs approximately 660 000 people or 2½ per cent of the population (1980). The average annual return on capital investment is only some 3 per cent (Checkley 1982), although this conceals marked variation in profitability. State support for agriculture was developed largely under the Agriculture Act 1947 to prevent a recurrence of the inter-war agricultural depressions and to improve investment levels (Ministry of Agriculture, Fisheries and Food 1979). The CAP, progressively adopted by Britain after 1973, was originally designed to increase agricultural productivity, to protect farmers' living standards, to stabilise markets and to ensure reasonable consumer prices (European Commission 1981). Unlike other Community funding most agricultural expenditure is mandatory and related to the three 'pillars' of the CAP: common prices in all member countries, preference in favour of Community production and uninhibited intra-community trade, and common financial solidarity. Finance for the CAP comes from the European Agricultural Guarantee and Guidance Fund (EAGGF, usually known by its French name of FEOGA) and from national governments. In 1979 FEOGA expenditure amounted to 73 per cent of the total Community budget.

This expenditure is largely used for agricultural prices support (Shoard 1980, Buckwell *et al.* 1981). First, a trade barrier exists around the Community with tariffs to raise the price of imports of what otherwise would be cheaper foodstuffs and to subsidise Community agricultural exports. Secondly, guaranteed intervention prices, above world prices, are paid to farmers inside the Community for all they can produce, thereby indirectly creating the famous surpluses. This system 'makes farmers the only group of producers who are guaranteed that whatever they produce will be sold at an acceptable price regardless of the level of demand and the overall level of production' (Shoard 1980, p. 24). Given this unlimited market at good prices the incentive is obvious for individual farmers to make improvements to their farms that can result in greater productivity at a reasonable rate of economic return. To quote one interpretation 'every undrained wetland represents forfeited profit' (Shoard 1980, p. 24). Furthermore, the investment necessary to make these productivity improvements is itself subsidised, in the form of grants for buildings, machinery and land improvement including drainage. Potential productivity gains for the individual farm and farmer are therefore not evaluated according to normal investment criteria, but on the basis of the economic return on that small contribution to total investment made by the individual entrepreneur.

The CAP also provides subsidies that encourage shifts in the type of farming. Given a large milk and milk-product surplus the FEOGA fund provides grants to encourage farmers to take cattle off the market. Annual CAP price rises – often opposed by the British Government – have also favoured cereals rather than animal husbandry (Body 1982), thus further encouraging the conversion of pasture land into arable cropping. The overall impact of these macroeconomic policies has been to encourage British farmers to increase cereal production, particularly wheat and barley, for which adequate land drainage is a necessary prerequisite. This policy has made Britain virtually self-sufficient in these crops despite the much lower production costs obtainable in the traditional corn-growing areas of the USA and Canada.

These policies have, of course, become intensely controversial. The National Farmers' Union and the Ministry argue that increased food production is desirable for a number of reasons. These include the strategic value of making Britain self-sufficient, improving Britain's balance of payments position and the contribution to world food production. Opponents, on the other hand (Shoard 1980, Body 1982, Bowers 1983), argue that the European Community suffers a surplus of most agricultural products, that there is therefore no need for increased food production, and therefore little need for further agricultural land drainage and the subsidised investment system which drives the process on. The opponents of land drainage also argue that increased British food production is unlikely in reality either to have any strategic value in modern warfare or to help those in Third World countries who face starvation, the latter because there is no incentive or satisfactory mechanism for distributing surpluses in this way. During 1983 and 1984 the agricultural industry became progressively threatened owing to increasing external economic and political pressures for change. This pressure centred on mounting government concern over the cost of financing the CAP which in turn began a complex process whereby the availability and level of subsidies became the subject of radical debate rather than taken for granted. Thus, whilst internally the agricultural institutions remain supremely strong politically and opposed to change, a major shift in the *status quo* – perhaps with the reduction or removal of grant aid for agricultural drainage – may be forced upon them by external factors beyond their immediate political control and related to changing central government needs and EEC policy.

Finance for flood protection and land drainage

An appreciation of the complexities of financing arrangements is crucial to an understanding of decisions concerning flood alleviation and agricultural drainage. Invariably these decisions, whether by individuals or organisations, are closely related to both the availability of finance and the conditions attached to its use. In addition, of course, the economic environment significant to these decisions is critically affected by Britain's agricultural support policies and thus principally by the CAP of the European Community discussed above.

The financing of land drainage and flood alleviation is supposedly based upon the fourth principle underlying land drainage law, also discussed above. This is that the cost of improvements should be financed by the section of the community which thereby benefits. The simplicity of this 'theory' is not matched in practice.

THE WATER AUTHORITIES

Water Authorities have various sources of income for land-drainage and flood-protection work (Fig. 2.4). Apart from relatively small government grants, income for revenue expenditure comes mainly from precepts on local authorities, principally the County Councils who in turn charge precepts on District Councils. Under Sections 45 to 47 of the Land Drainage Act 1976 the Water Authorities have powers to demand these contributions from the County Councils for land drainage costs and they are intended to reflect the benefits of land drainage to the local community. Land drainage costs are apportioned amongst the County Councils within a Water Authority area on the basis of the 'penny rate product' for their areas (the product of a rate of one penny in the pound). This is intended to equalise the burden of charges between Councils with high and low rate income. A much smaller proportion of Water Authority income for land drainage comes from precepts on Internal Drainage Boards. Again these precepts are intended to reflect the benefits to Internal Drainage Districts of the Water Authorities' land drainage work on main rivers and sea defences.

Water Authorities also have powers to levy general drainage charges: 'other income' in Figure 2.5. These charges are used extensively by the Anglian Water Authority, but not at all by Severn Trent, and can be raised

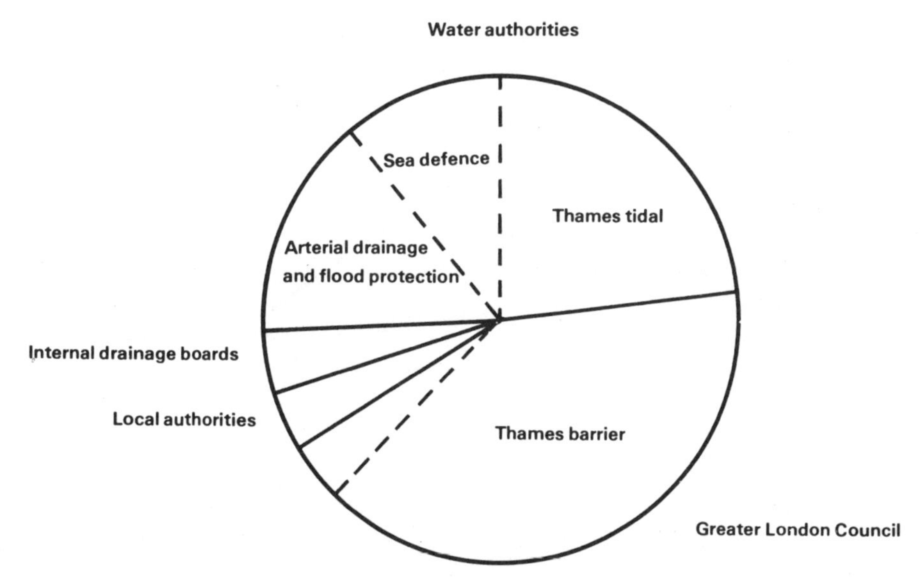

Figure 2.4 Responsibilities for capital expenditure on land drainage (from National Water Council 1982).

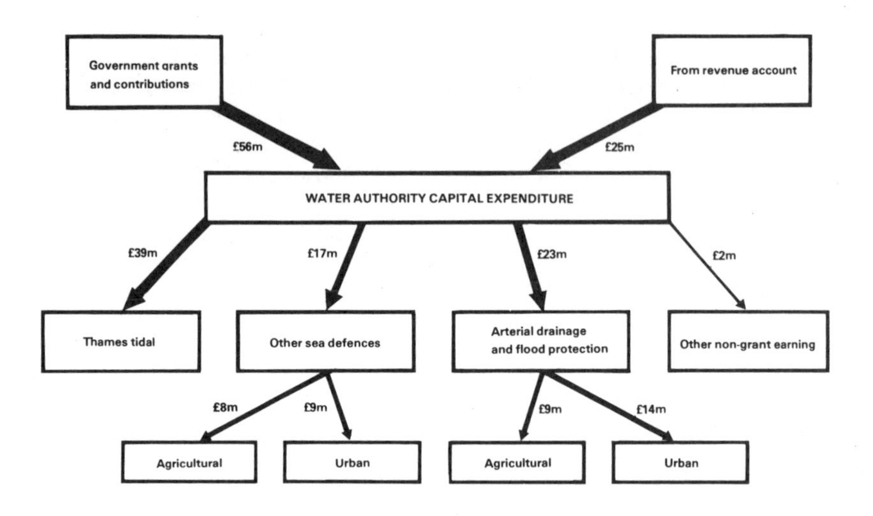

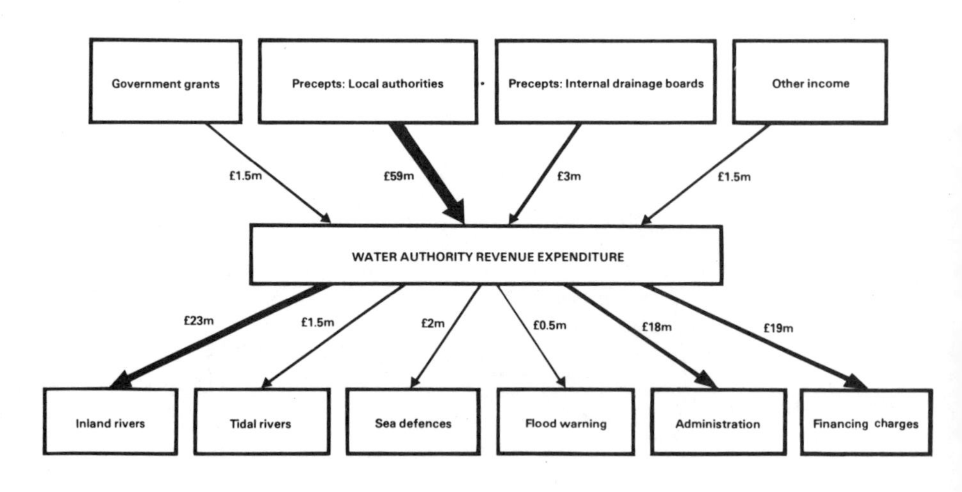

Figure 2.5 Water Authority expenditure on land drainage 1980–1 (from National Water Council 1982).

from the owners of all agricultural land within a local land drainage district which lies outside of an Internal Drainage district (Fig. 2.3). The charges are based on rateable value, but agricultural land was absolved from rate payments to local authorities (i.e. local taxes) in 1929, during a period of intense agricultural depression, and the rating has never been reimposed. The Land Drainage Act 1976 therefore prescribes a procedure designed to ensure that the amount charged is equivalent to the charge that would be paid on the land if it were rated! Special drainage charges can also be levied in specific areas outside Internal Drainage Districts where it appears that drainage works would advantage their agriculture.

Having received contributions from all these sources Water Authority revenue expenditure on land drainage was about £64 million in 1980–81. Of this the major sum was spent on river maintenance, £18 million on administrative, research and technical support, £19 million on interest and depreciation and the rest on maintaining sea defences and flood warning systems.

The major responsibility for capital expenditure on land drainage rests with the Water Authorities (Fig. 2.4). This expenditure is financed by loans and government grant aid (Fig. 2.5), although self-financing from surplus revenue income over expenditure is increasing (Penning-Rowsell & Parker 1983). Sources of borrowing are limited mainly to the National Loans Fund, with repayment periods of between 5 and 25 years for new capital works, within ceilings strictly controlled by central government. However, foreign borrowing is also possible although only with central government approval. Of the Water Authorities' 1980–81 capital expenditure on arterial drainage and flood alleviation over one-third goes on improving agricultural land and under two-thirds on protecting existing urban development. Capital expenditure on sea defence ran at a high level in 1980–81 owing to the costs of improving the Thames tidal defences downstream of the Woolwich barrier (Fig. 2.4).

Unlike other water services land drainage qualifies for government grant aid (Section 90 of the Land Drainage Act 1976). The Ministry of Agriculture, Fisheries and Food or the Welsh Office sets a grant ceiling for each Water Authority and rates of grant vary between Authorities according to a complex formula intended to reflect the magnitude of their land drainage problems and the rate income of the relevant County Councils. Thus, for example, the level of grant in 1977–9 was 40 per cent in the Severn catchment, 64 per cent for the Norfolk and Suffolk areas of the Anglian Water Authority but only 11 per cent in Northumbria. An extra 15 per cent grant is given for tidal and sea defence works. Capital expenditure by Water Authorities on arterial land drainage and sea defence works – together with the grant-aid contribution – is shown for 1981–2 in Figure 2.6.

There are two further sources of grant for Water Authority land-drainage work. One is the European Community's FEOGA Fund, which can contribute directly to schemes, such as substantial arterial drainage schemes, that promote the objectives of the Common Agricultural Policy. The Fund contributes a maximum of 25 per cent of the total cost so long as the Member State approves the project for a national grant. Drainage

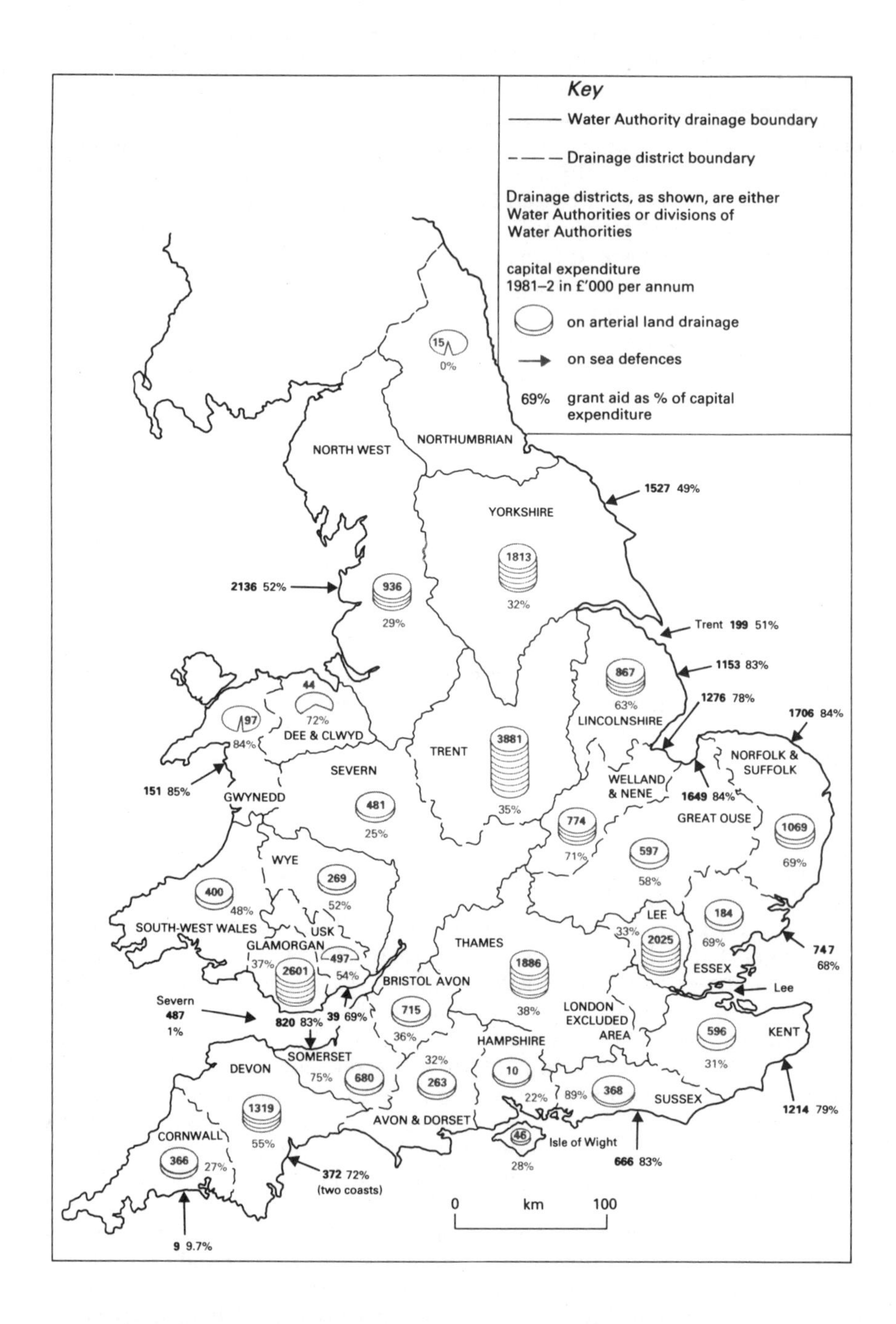

Figure 2.6 Capital expenditure on land drainage and sea defence by drainage districts 1981–2 (data from Ministry of Agriculture, Fisheries and Food).

authorities that will benefit are expected to provide at least 20 per cent of the cost. Grants are also available for British Development and Special Development areas from the Community's Regional Development Fund.

LOCAL AUTHORITIES

Local authority finance for land drainage and flood protection is a thorny issue. County Councils' land drainage capital expenditure can be financed from loans from central government within a government set loan ceiling.

All County Councils receive annually from the government a global loan sanction to cover all the investment schemes which the County or District Councils wish to undertake, including land drainage and flood alleviation works. The method of distribution of the loan allocations varies but a common method is for the County to retain 50 per cent and the rest to be distributed amongst District Councils according to their populations. Whereas District Councils could previously borrow the costs of a land drainage scheme the Land Drainage Act 1976 (Section 104) appears to have removed this borrowing power (Wood 1980). Where large schemes are concerned, therefore, the District Councils must rely upon their allocation via the County Council but this is frequently inadequate to meet scheme costs.

Under Section 91 of the Land Drainage Act 1976 grants can be paid by the Ministry of Agriculture, Fisheries and Food to District Councils as well as to County and Metropolitan Councils. The grants payable are variable up to a maximum of 50 per cent of the cost and are based upon the Councils' previous capital expenditure. Local authorities can also take advantage of the European assistance discussed above.

INTERNAL DRAINAGE BOARDS

Internal Drainage Boards' income is from drainage rates under Sections 63–9 of the Land Drainage Act 1976 (see Fig. 2.4). The receipts are divided between owners' and occupiers' rates. The owners' rate is intended to cover the cost of new drainage work or the improvement of existing works and any payments from the Board to a Water Authority. The occupiers' rate is to cover all other expenses notably on maintenance and administration. Rates are again determined according to what would be the rateable value of the land in question, as with general drainage charges. Different rates may be set on the grounds of differential benefit from drainage and ratepayers may petition Boards to make a differential rating order. Grant aid of up to 50 per cent of the total cost of a capital scheme is also available to these Boards from the Ministry of Agriculture, Fisheries and Food.

FIELD DRAINAGE FINANCE FOR LANDOWNERS

A variety of grants is available to landowners and farmers to enable them to meet the expense of improving their land and thereby increasing their contribution to national food production. These improvement grants contribute to the cost of field drainage and are an important source of capital for landowners and farmers. Under the Agriculture and Horticulture Grant Scheme (AHGS) the Ministry offers 'Investment Grants' on capital investment (Ministry of Agriculture, Fisheries and Food 1981a, 1982). These grants are available for the provision, replacement or improvement of field drainage, including underdrainage and ditching, as well as for works to prevent the flooding of agricultural land. For field drainage the rates of grant vary from 37.5 per cent (standard rate) to 70 per cent (in less favoured areas), and from 22.5 per cent (standard rate) to 50 per cent (in less favoured areas) for flood-protection works.

A second Ministry fund – under the Agriculture and Horticulture Development Scheme (AHDS) – provides 'Development Grants'. Like AHGS grants they are available for field drainage and flood prevention works where these are part of an approved development plan (Ministry of Agriculture, Fisheries and Food 1980). However, higher rates of grant are available under AHDS: 50 per cent (standard rate) to 70 per cent (in less favoured areas) for field drainage and 32.5 per cent (standard rate) to 50 per cent (in less favoured areas) for flood-prevention works.

Under an administrative charge in 1980, designed to reduce civil servant numbers, the local Ministry staff no longer have to give prior design advice and approval to farmers to undertake field drainage works (Shoard 1980, p. 28). Instead the farmer pays for the necessary works and reclaims the grant, thus breaking a significant link in the field-to-coast chain integrating the management of land drainage.

ORGANISATIONAL STRUCTURES AND FINANCE

Combining the discussions above concerning the economic, financial and administrative context of flood alleviation and agricultural drainage we can see clearly that it is 'agriculturalists' who ultimately both control the 'purse strings' and use the finance. 'Agriculturalists' argue that this is appropriate and fair since they have the strongest interest in land drainage. Alternatively, it can be argued that the mutually supporting and powerful community network of agricultural interests – of which the Land Drainage Committees and government departments are the crucial parts – is undesirable since other legitimate interests are disadvantaged. Urban drainage and nature conservation interests are particularly strong in making such claims.

The financial arrangements have also been criticised for 'featherbedding farmers' (Wakelin 1980). This is because the fourth principle underlying land drainage law, concerning the beneficiaries of drainage improvements paying accordingly, appears to be compromised. Farmers take advantage of tax payers' money through Ministry drainage grants. They also benefit significantly from grant-aided arterial drainage improvements. Thus it is the British tax payer who pays for an important proportion of agricultural drainage improvements and thereby helps to raise the profits of farming drained land (Parker & Penning-Rowsell 1980).

In defence of the existing system agriculturalists argue that encouraging the farming industry to grow more food and to reduce food imports all helps to increase Britain's self-sufficiency. Thus, it is argued that the prime reason for grant-aiding agricultural drainage improvements is to promote the national interest. Parallel grant aiding of urban flood hazard alleviation and sea defence schemes occurs for indeterminate reasons related to welfare concepts of the community protecting arbitrarily disadvantaged minorities. One must conclude, however, that the clear principles underlying the law become fuzzy in their application.

Promoting, financing and executing an agricultural drainage or flood alleviation scheme

The administrative system for promoting, financing and executing agricultural drainage and flood alleviation schemes varies considerably according to the particular case and the drainage authority or person promoting the scheme. Within broad bounds, however, the process can be generalised as a set of procedures that will condition the sequence with which decisions can be made. These procedures reflect, however, the general tendency in Britain for flood alleviation schemes to be structural in character rather than non-structural alternatives.

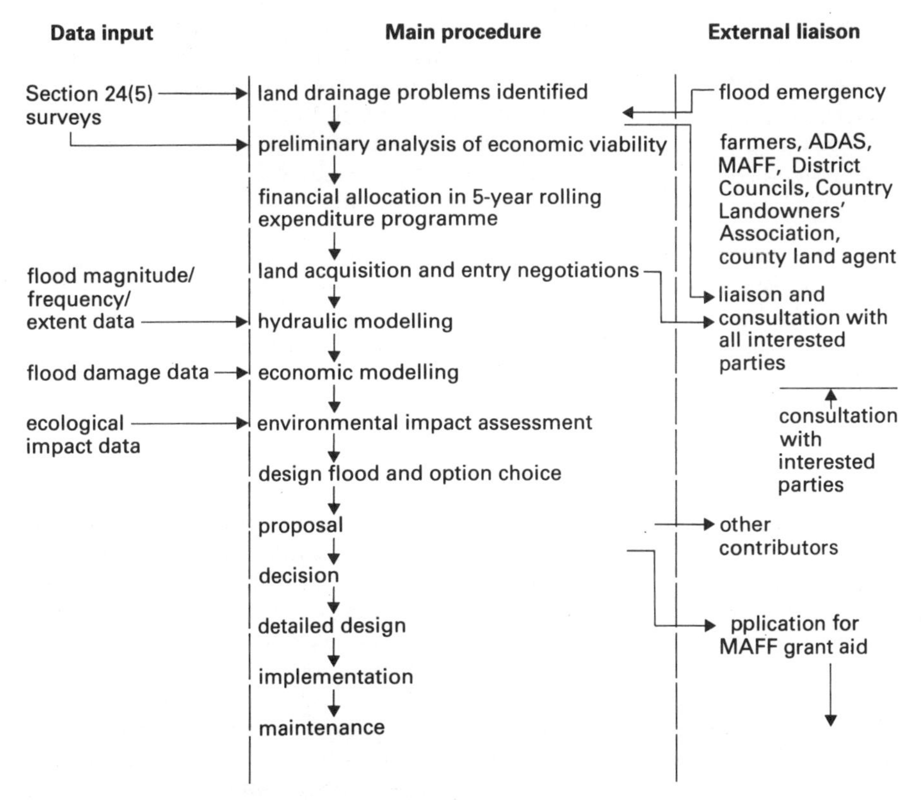

Figure 2.7 Simplified administrative procedure for promoting a Water Authority flood alleviation or land drainage scheme.

By far the largest expenditure is made by Water Authorities and the main stages in the administrative procedure can be demonstrated using an idealised example (Fig. 2.7). Drainage problems are traditionally identified by flood emergencies or by farmers, or perhaps by the Ministry notifying Water Authorities or Land Drainage Committee members of drainage needs. Under the Water Act 1973 Water Authorities have had to undertake Section 24(5) land drainage surveys to identify drainage and flood problems and these surveys provide an additional data source and a potentially more systematic basis for planning land drainage expenditure (Ch. 3).

Viable drainage or flood alleviation schemes are usually identified by preliminary subjective analysis of the economics of remedial measures for particular problems. With the agreement of the Regional Land Drainage Committee the potentially viable schemes are then included in the Water Authority's rolling five-year capital expenditure programme by allocating a capital sum to prospective schemes.

The decision to include a scheme in the expenditure programme will also normally follow wide consultations. This is often through the members of the Land Drainage Committee or through Water Authority officers or members. A crucial stage in the promotion of a scheme is also the acquisition of land and the negotiation of entry agreements where these are necessary and involve 'imperative consultations' (Ch. 6).

Detailed feasibility analysis and design of a flood-alleviation scheme – whether structural or non-structural – requires the modelling of floods of a range of magnitudes and frequencies, routing these through the river sections concerned and establishing the extent of flooding for floods of different magnitudes (Ch. 3). Engineering consultants or Water Authority staff may perform these hydrological, hydraulic and economic feasibility studies. Establishing the economic viability of scheme options is crucial if capital is to be used wisely and in Britain the Ministry of Agriculture, Fisheries and Food grant aid is only forthcoming for schemes with a favourable benefit–cost ratio (Ch. 4). Since flood alleviation and agricultural drainage schemes may have adverse environmental effects (Ch. 5), the impact of proposals in this respect also requires assessment, through consultation with interested and affected parties and by interagency negotiation prior to scheme selection (Ch. 6).

Scheme proposals need not be submitted for Ministry grant aid but virtually all flood alleviation and most agricultural drainage schemes seek some measure of government financial aid. Important beneficiaries may also be asked to contribute to scheme costs in line with the principles underlying land drainage law. Once a decision is made on the technical efficiency and economic viability of the scheme, by both the Regional Land Drainage Committee and the Ministry, a decision to proceed to the detailed design stage usually follows. Again consultants may be involved and outside contractors will probably undertake any structural works although Water Authorities generally use their own 'direct labour' for maintenance. Public consultation is required through all these stages so that detrimental impacts on individuals and groups may be foreseen and reduced.

The Water Authorities' administrative procedures for promoting schemes – here idealised as a neat and logical linear sequence – prompts several observations. In practice the capital sums allocated to a scheme or programme of schemes early in the procedure may become immutable without 'loss of face' or threats to engineers' job security if more detailed engineering, economic and environmental feasibility studies later suggest reductions in the programme budget. The system of annual budgeting, as with all government expenditure, can result in over-hasty 'planning' towards the end of financial years as money unspent cannot generally be carried over to the next year. On the other hand, delays within the complex multilayered administrative system are common. The budget for land drainage as a whole is consequently often underspent and, more significantly, flood problems can persist for years as administrative negotiations continue.

Preliminary scheme analysis and consultation could be based as much on environmental impact as on economic grounds and this is now happening. However, consultation raises a variety of problems. These include identifying all interested parties, genuinely involving these at an early enough stage to allow incorporation of their information and views, and getting binding agreement amongst the many interests involved. Water Authority staff or Land Drainage Committee members still unfortunately get committed or 'locked in' to a scheme before adequate economic, environmental or political evaluation. This remains a principal cause of the considerable antagonism surrounding some flood alleviation and agricultural drainage schemes.

Assessment

Land drainage law and organisational structure is the consequence of policies developed by politicians and civil servants acting within the influential and constantly evolving 'force field' of the wider political economy. Similarly the financial incentives for agricultural drainage are the result of policies that reflect decisions made within this wider environment, the importance of which is underscored by recent trends threatening the land drainage *status quo* over grant aid and which, under the Water Act 1983, perhaps prepare the way for at least partial water industry privatisation. At a more routine technical level decisions are continually being made concerning whether to proceed with particular schemes, about standards and means of protection and about how to finance the operation and with whom to consult.

Policies and decisions are ultimately fixed by individuals or small groups and Lindblom's (1959) notion of disjointed incrementalism appears the most accurate description of this decision making. This in turn would appear to emphasise the rôle of the individual making decisions based on individual whim. However, the forces acting upon decision-makers are all important. This is not to say that individual cognition and behaviour are unimportant, but that policies and decisions can only be fully appreciated in relation to their institutional context which introduces the influence of administrative, economic and political forces (Fig. 1.6). Thus a number of the institutional factors and forces significant to our understanding of policy and decision making have so far been revealed.

The first is that flood hazard alleviation in Britain is deeply rooted within a 'land drainage' legal context. Unlike the situation in the USA or Australia it cannot be understood without a full appreciation of this 'agricultural' framework. The agricultural emphasis deeply pervades all institutional arrangements for flood alleviation including the law, organisational structures, economic context and financing arrangements, which are only marginally different between Scotland and England and Wales.

There is no doubt, secondly, that Britain's land-drainage and flood-alleviation institutions are complex and powers are remarkably fragmented. This has led to demands for institutional reform (Wilkins 1980). Confusion about the rôle of different organisations is not uncommon amongst civil servants, planners, land-drainage engineers and especially amongst the general public. The arrangements have been criticised as 'a chaos of authorities with an absence of authority' (Wilkins 1980). Nevertheless the organisational and legal arrangements have arisen to meet clear objectives including local control, subsidised hazard relief and agricultural investment, the protection of individuals' private property rights and permissive rather than mandatory powers. Many of these objectives can become incompatible, but if they are to be met – and this requires a value judgement – then something akin to the current arrangements is necessary.

Thirdly, the economic and political forces favouring the *status quo* or change only at the margin of the *status quo* are powerful. The current institutional system for both flood alleviation and agricultural drainage primarily benefits farmers and landowners. It is therefore in their interests to maintain this *status quo*. The forces favouring this situation are strengthened by the characteristics of the other relevant organisations. Amongst these characteristics is the 'survival' objective which in Britain appears to militate against change and maintain the prevailing conditions without which the dominant professional groups and interests may be threatened. Large organisations appear to grow inherently more conservative both as to procedures and to objectives (Brooks 1974). Defensiveness and conservatism are features not just of the Ministry of Agriculture, Fisheries and Food or the Water Authorities. They are traits that appear well developed in all British government agencies, the civil services as a whole and even Polytechnics. This means that radical institutional reform, no matter how desirable, becomes progressively more unlikely (O'Riordan 1981) unless imposed upon institutions by external economic and political forces.

Fourthly, decisions are clearly the outcome of a complex power struggle between competing groups and conflicting viewpoints about the use of land, finance and the impact of decisions upon different interests. This analysis would support a pluralist view of society which emphasises the healthy conflict between interests competing for the same resources. However, the rôle of macroeconomic forces is also critical. The agricultural policies of the Economic Community – and British agricultural protectionism before 1973 – has created an environment in which individual farmers and landowners respond directly to the incentives towards increased production for which drainage may be necessary or simply more profitable. National economic conditions, rather than just pressure-group politics, condition the overall expenditure ceilings against which urban flood alleviation is planned; the government rôle in decision making ensures that central control is maintained.

Our continuing analysis in subsequent chapters may well appear to emphasise the details of decision making by those 'close to the ground'. However, this should not be taken to imply the insignificance of the wider institutional context. Nor should it imply that hydrological and economic analysis necessarily leads to an engineering structural solution to hazard reduction and agricultural improvement. It does, however, continue to emphasise the problems of policy formulation and decision making in this complex field.

3 *The hydrologic, hydraulic and hydrographic analysis of flooding and land drainage problems*

Introduction: the developing data base

An essential prerequisite for any flood hazard reduction programme, including the design of structural flood alleviation works or non-structural alternatives, is adequate information to define the problem. Such information must include knowledge of the areas subject to flooding and impeded land drainage. Hydrological and hydrometeorological records are also essential for the statistical analysis of floods. Data on soil characteristics and fluctuations in the water table are central to the analysis of waterlogging. Frequently, however, such information has been lacking in the past, either because of limited instrumentation and inadequate surveys, or because the available records are short or incomplete and hence difficult to use for the prediction of likely future flooding or agricultural drainage problems.

Major efforts over the last 30 years in the field of hydrometry have considerably improved the data base in Britain. Many new river gauges have been installed, particularly following the Water Resources Act 1963. Automated measuring methods, aerial photography and even satellite imagery have all greatly facilitated the recording of flood events. Land-drainage surveys have also systematically identified the location of flood hazards and land drainage problem sites following the requirements for these surveys under Section 24(5) of the Water Act 1973. Historical flood records have been investigated and attempts made to incorporate these data, which are often qualitative in nature, into flood analysis to augment those obtained from hydrometric surveys (Potter 1978).

Data alone, however, are of limited value. The development of satisfactory analytical techniques has long been recognised as being vital. As Howe *et al.* (1967) observed, 'the history of the study of floods is essentially the record of a search for adequate tools to analyse past hydrological events in terms of future possibilities of occurrence'. Similarly, the report on *Flood studies for the United Kingdom* (Institution of Civil Engineers 1967) emphasises that 'one of the most difficult problems in flood hydrology is the interpretation of recorded events to derive estimates of the future probabilities of events of different sizes'. Just as the raw hydrological data has been improved in recent years so has research in Britain and elsewhere provided a wider range of more sophisticated analytical techniques.

Research on flood magnitude and frequency

As physical phenomena, in contrast to their hazardous effects, floods have been described as both three-dimensional and binary (Howe *et al.* 1967).

Floods are binary in that they are most influenced by two sets of factors: climatic and physiographic. Various studies have been undertaken to identify the relationships between these factors and flood characteristics, many involving multiple regression techniques designed to isolate the contribution different parameters make to peak flows (Potter 1961, Benson 1962, Rodda 1969). The three-dimensional nature of floods derives from their three significant characteristics, namely magnitude, frequency and timing. Much flood hydrology is concerned with the interaction between these characteristics.

Magnitude/frequency studies are of vital importance in hydrological analysis (Natural Environment Research Council 1975); they also are fundamental to many areas of flood hazard research and the economic evaluation of flood alleviation schemes. Timing studies include hydrograph analysis (Weyman 1975) and flood routing (Lawler 1964). As Newson (1975) has indicated, there are situations when engineers need to predict the whole hydrograph rather than just the flood peak. This can occur when river flow records are short and longer rainfall records are used in the simulation of large floods using a hydrograph model and rainfall analysis techniques, such as probable maximum precipitation, when knowledge of flood duration is needed. Knowledge of the whole hydrograph is also needed for real-time forecasting of floods using instantaneously recorded data on catchment rainfall, perhaps from weather radar and linked to flood warning systems (Newson 1975). Knowledge of flood volumes, rather than just the size of peaks, is also needed when designing reservoir storage for water supply or as a flood-alleviation strategy.

EARLY WORK

An important analysis of flood hydrology research before 1965 is provided by Wolf (1966) who reviews both the early magnitude studies and the subsequent frequency investigations.

Attempts were made in the early 20th century to establish predictive flood magnitude relationships using relatively simple formulae. The fastest developments occurred in relation to the design of urban storm sewers and many formulae from early analyses relate peak flows to catchment area. Probably the most widely used in Britain is from the Institution of Civil Engineers' 1933 Committee Report. While briefly considering flood frequency, and the effects of catchment character and precipitation on the flood hydrograph, this report concluded that 'the only acceptable method for presenting flood information consisted of plotting recorded instantaneous peak flows against the contributing surface catchment areas' (Institution of Civil Engineers 1933). The Institution's report used the common formula:

$$Q_p = c.A^m$$

where Q_p is the maximum discharge produced by a catchment of area A, c is a constant and m an exponent. The result of the Institution's study was a 'Normal Maximum Flood' line for upland catchments, with a formula

$$Q = 1000.\ A^{0.6}$$

with discharge Q in cusecs (cubic feet per second) and area A in square miles.

Nevertheless, peaks higher than the Institution's Normal Maximum Flood line were subsequently experienced (Learmonth 1950, Smith 1965, Chapman & Buchanan 1966). The report was therefore updated and republished in 1960 (Institution of Civil Engineers 1960). Hall (1980) discusses other studies which relate flood magnitude to catchment area using alternative formulae, including the Craeger curves (Craeger 1945). Chow (1964) reported that frequency analysis of storm events and peak discharges was recognised as being useful following the work of Herschel and Freeman in the 1880s. Modern statistical analysis of flood magnitude/frequency relations dates from the work of Gumbel (1941, 1958) who made fundamental theoretical contributions to the understanding of extreme value distributions of which floods are but one example. A further important contribution was made by Powell (1943) who modified and developed plotting paper for use in magnitude/frequency studies seeking to predict major flood peaks from short records.

Flood frequency studies in the USA developed rapidly after the work of Gumbel and Powell, principally in the 1950s through the United States Geological Survey. Dalrymple (1960) developed a technique of regional flood frequency analysis which involved the derivation of two curves. One shows an average frequency curve defined for the region, which indicates the ratio of a flood to the mean annual flood. The second curve relates the mean annual flood to catchment area. The method is important in allowing the derivation of a frequency curve for an ungauged site and a detailed application of the method in Britain can be found in Howe *et al.* (1967).

In addition to the statistical flood estimation approaches, described above, deterministic flood estimation methods have provided important techniques for deriving the whole flood hydrograph. Such methods 'determine' likely future flood magnitude through predictions from its relationship with inputs such as rainfall and catchment characteristics, both of which affect flood volumes and timing (Ward 1978, p. 71). A highly important technique which has been widely used in flood studies is the unit hydrograph (Snyder, 1938, Nash 1960, Chow 1964, Cordery 1971). The basis of the method is the finding that this typical or unit hydrograph encapsulates many of the physical characteristics of the catchment area. Once this hydrograph has been determined, therefore, it can be used to predict flow patterns for the catchment for rainfall of any duration or intensity.

THE *FLOOD STUDIES REPORT*

Continual flood problems throughout the 1960s in Britain prompted a major hydrological research effort resulting in the *Flood studies report* (Natural

Environment Research Council 1975). The aim was to improve the methods for estimating flood frequency and magnitude which are used in design situations. The origins of the research programme are described in the report *Flood studies for the United Kingdom* (Institution of Civil Engineers 1967):

> So few of the existing improved techniques in use overseas have been adopted and applied to British conditions that their introduction alone would quickly result in major improvements in British flood hydrology.

The study was sponsored by the Natural Environment Research Council, backed by the major consulting engineering firms and the Ministry of Agriculture, Fisheries and Food, and was undertaken at the Institute of Hydrology in the early 1970s. The Flood Studies team followed closely the recommendations of the 1967 report of the Institution of Civil Engineers in its approach to hydrological estimation. In summary (Sutcliffe 1978, p. 43), these were that the investigations should:

> Examine all aspects of flood hydrology; meteorological records should be studied to understand the causes of floods, to extend flood records, to assist in flood frequency analyses and to provide estimates of probable maximum precipitation. All available flood records should be assembled and reviewed; frequency analyses and correlations with catchment characteristics should be undertaken to improve single station frequency distributions and to estimate flood frequencies at ungauged sites; unit hydrographs, soil infiltration characteristics and snow-melt should be studied to derive precipitation/runoff models for use with the results of meteorological studies; flood routing techniques should be reviewed and tested.

An important contribution of the *Flood Studies Report* to flood research in Britain, and extending also to Northern Ireland, was the major advance in collating the hydrometeorological data base. Records were collected for all parts of the country and the result is a major new source of flood information (Natural Environment Research Council 1975, vol. 4). The hydrometeorological studies were based on analyses of records from more than 600 daily rainfall stations with an average of 60 years of record supplemented by daily and monthly records from 6000 stations for the period 1961–70. All available flood flow records were also collected and information from 533 sites was used which, in total, yielded some 6000 station-years of records.

As Rodda *et al.* (1976) have indicated, volume 4 of the *Flood Studies Report* has therefore made available most of the flood flow information for Britain and the volume can be used to obtain flow data for most parts of the country. Mean daily and peak flow data are retained by the Water Authorities, but the main source of flood data remains the Institute of Hydrology, Wallingford, with their constantly updated Flood Data Archive (Penning-Rowsell & Parker 1984).

The *Flood Studies Report* contained much theoretical discussion of techniques before proposing specific methods for frequency and magnitude estimation. The report is founded on the techniques of flood frequency and unit hydrograph analysis and indicates that there are two main routes to obtaining the estimate of the design flood. There is, first, a statistical approach involving flood frequency analysis and, secondly, a deterministic approach which can be applied through unit hydrograph analysis involving synthesis of the flood corresponding to a design storm. Sutcliffe's (1978) *Guide to the Flood Studies Report* indicates in summary form which method should be used, suggests when and how to mix the techniques, and usefully provides a hybrid approach with a simple preliminary technique for deriving the estimated maximum flood discharge.

Sutcliffe (1978, p. 2) also indicates that the choice between the statistical and deterministic approaches largely depends on the answer to the following questions:

(i) Is the hydrograph, or the detailed shape of the flood, required in addition to the instantaneous peak flow (as for example when the flood needs routing through a reservoir)?
(ii) Is an estimate of the maximum flood required, rather than an estimate of the flood of a given frequency or return period?

If the answer to either question is positive, this indicates that the unit hydrograph method is required. However, if only an estimate of the maximum flood of a given return period is required, then several methods are available. The choice of which of these to use will depend on a number of factors including the availability of rainfall and runoff records, their reliability and length. In many cases individual judgement is required, but a useful flow diagram indicates how the choice of method can be made (Fig. 3.1).

In the case of the statistical approach the theoretical advantages of annual maximum series and partial duration (peaks over threshold – POT) series are discussed and the form of the distribution to be used with each series is indicated. The general extreme value distribution is suggested for use with the annual maximum series, whereas the exponential distribution should be used with the partial duration series. Plotting positions using various types of plotting paper are outlined, and the derivation of the regional curve is explained as examining the annual series jointly at several stations.

Estimation of the mean annual flood is crucial to all the statistical techniques. The report allows the calculation of the mean annual flood from catchment characteristics to apply the method to ungauged sites. An average country-wide equation was established of the form:

$$\bar{Q}=0.0201\ \mathrm{AREA}^{0.94}\ \mathrm{STMFRQ}^{0.27}\ \mathrm{S1085}^{0.16}\ \mathrm{SOIL}^{1.23}\ \mathrm{RSMD}^{1.03}\ (1+\mathrm{LAKE})^{-0.85}$$

where $\bar{Q}$ is the mean annual flood, AREA is area in km^2, STMFRQ is stream frequency in junctions/km^2, S1085 is stream slope in m/km, SOIL is a soil index (see report or guide Appendix A), RSMD is net 1-day rainfall of 5-year return period in mm, LAKE is a lake index (see report or guide Appendix A).

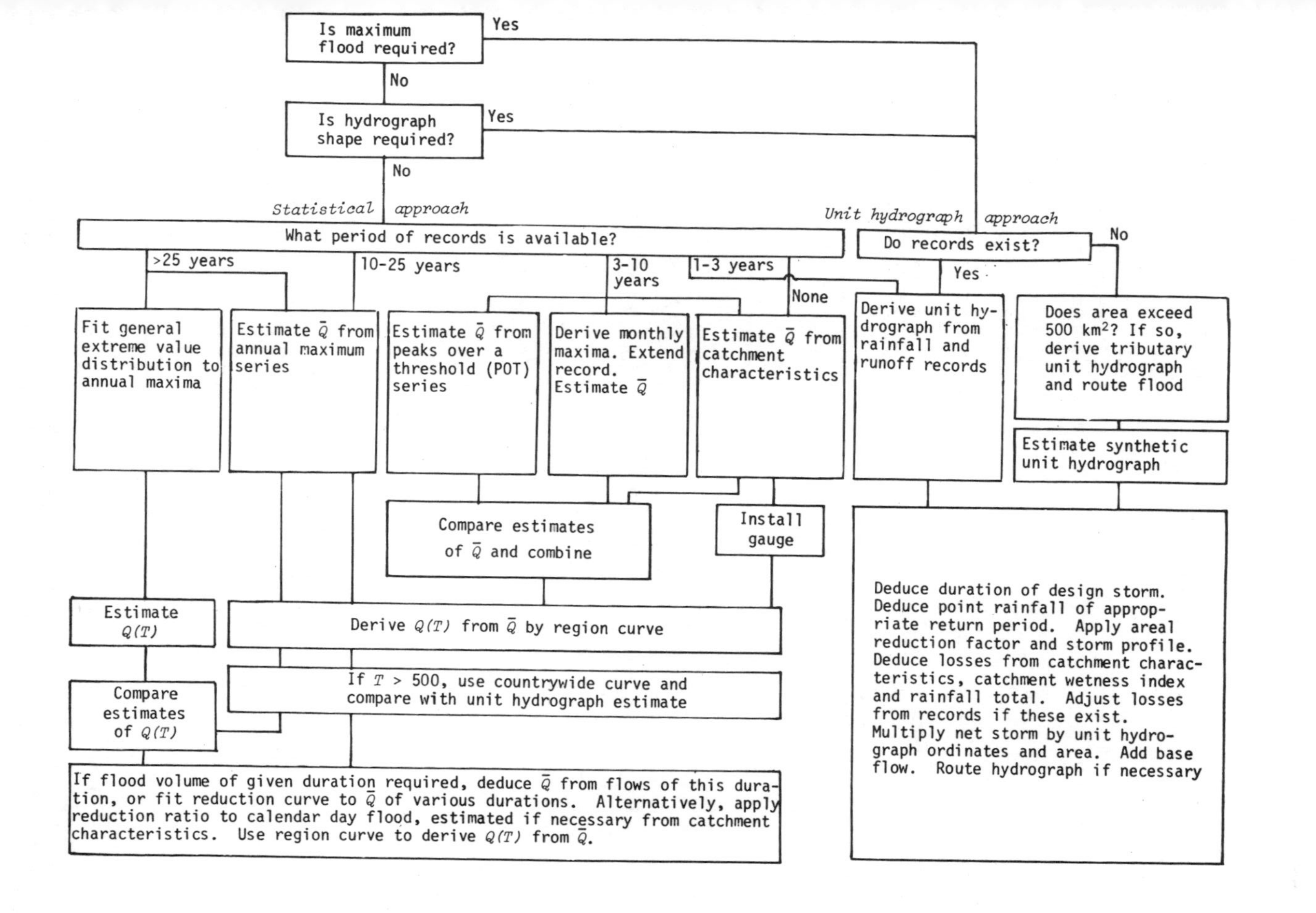

Figure 3.1 Procedure for the estimation of the design flood (from Natural Environment Research Council 1975, Sutcliffe 1978).

Table 3.1 *Flood Studies Report* regional curve ordinates (from Sutcliffe 1978).

Region	Hydrometric areas	Return period (years)						
		2	5	10	25	50	100	200
1 Northern Scotland	1–16, 88–97, 104–108	0.90	1.20	1.45	1.81	2.12	2.48	2.89
2 Southern Scotland	17–21, 77–87	0.91	1.11	1.42	1.81	2.17	2.63	3.18
3 North-east England	22–27	0.94	1.25	1.45	1.70	1.90	2.08	2.27
4 Severn/Trent	28, 54	0.89	1.23	1.49	1.87	2.20	2.57	2.98
5 East Anglia	29–35	0.89	1.29	1.65	2.25	2.83	3.56	4.46
6/7 Thames/Essex/ Southern England	36–44, 101	0.88	1.28	1.62	2.14	2.62	3.19	3.86
8 South-west England	45–53	0.88	1.23	1.49	1.84	2.12	2.42	2.74
9 Wales	55–67, 102	0.93	1.21	1.42	1.71	1.94	2.18	2.45
10 North-west England	68–76	0.93	1.19	1.38	1.64	1.85	2.08	2.32
Great Britain (average)		0.89	1.22	1.48	1.88	2.22	2.61	3.06
Ireland		0.95	1.20	1.37	1.60	1.77	1.96	2.14

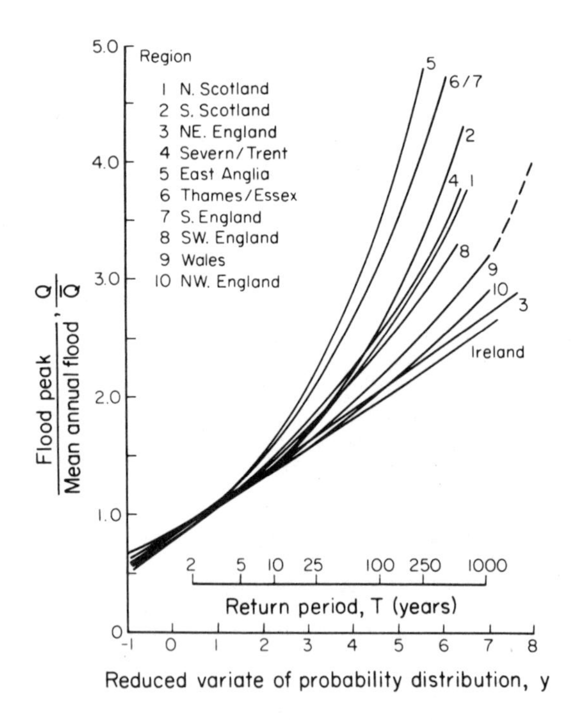

Figure 3.2 *Flood Studies Report* regional curves showing average distribution of $Q/\bar{Q}$ in each region (from Rodda *et al.* 1976, Natural Environment Research Council 1975).

Regional multipliers replace the constant 0.0201 in the above equation to provide an improved estimate. Estimation of the mean annual flood by extension of short records is also possible. Techniques are also available for the assessment of $\bar{Q}$ from partial duration series, from annual maximum series, and for floods of various durations. The ordinates of the regional frequency curves are shown in Table 3.1 and Figure 3.2 and Q (T) can be estimated from the mean annual flood either from flow records or, in the case of ungauged catchments, from catchment characteristics and then multiplying $\bar{Q}$ by the appropriate value of $Q/\bar{Q}$ derived from the regional curve. Clearly where long records exist it is possible to estimate Q (T) directly from the annual maximum flow series (Sutcliffe 1978, p. 15).

The recommended unit hydrograph approach is more complex than the statistical method (Sutcliffe 1978, p. 15). Derivation of the unit hydrograph can be from either rainfall and runoff records or from catchment characteristics if no record exists. Many of the procedures follow conventional unit hydrograph practice. Once the catchment rainfall total and a time profile is provided the proportion of the design storm providing immediate runoff is calculated from an equation that includes soil characteristics, an appropriate index of antecedent conditions, and the rainfall total. The ordinates of the unit hydrograph, once derived, are then used to multiply the net storm rainfall to provide the design flood with base flow added. A simulation procedure is also available to help where records are short. If adequate records are available, a minimum of five hydrographs is recommended for derivation of the unit hydrograph for the design site, and observations on the choice of hydrographs are presented. Sutcliffe (1978, p. 19) provides a flow chart showing the steps in the estimation of a design hydrograph using the unit hydrograph method (Fig. 3.3). Each step is explained by means of a worked example and the estimation of the maximum flood for a situation without rainfall and runoff data from the catchment or nearby catchments is outlined.

One further aspect of the *Flood Studies Report* concerns flood routing. This is an essential element in many flood studies and it involves attempts to assess the attenuation of the flood hydrograph with distance downstream. Given the assumptions inherent in the unit hydrograph method of spatially uniform rainfall the method is not normally used on large catchments (Newson 1975). Flood routing is used instead and various methods, including the 'Muskingum' method (McCarthy 1938, Cunge 1969) and the diffusion method (Hayami 1951, Hayashi 1969, Price 1973) have been applied to routing problems.

The *Flood Studies Report* also examines the problems of overbank flow, which can cause considerable variation in the parameters used in flood-routing methods from those that apply in the 'in bank' situation for the same river. A new flood-routing method is presented using a parameter that allows the speed of the flood wave to vary with discharge, and a second parameter which describes the effect on a flood wave of the irregularities in the channel geometry (Sutcliffe 1978, p. 48).

DEVELOPMENTS SINCE THE *FLOOD STUDIES REPORT*

Methods of flood routing have been developed further in studies at the

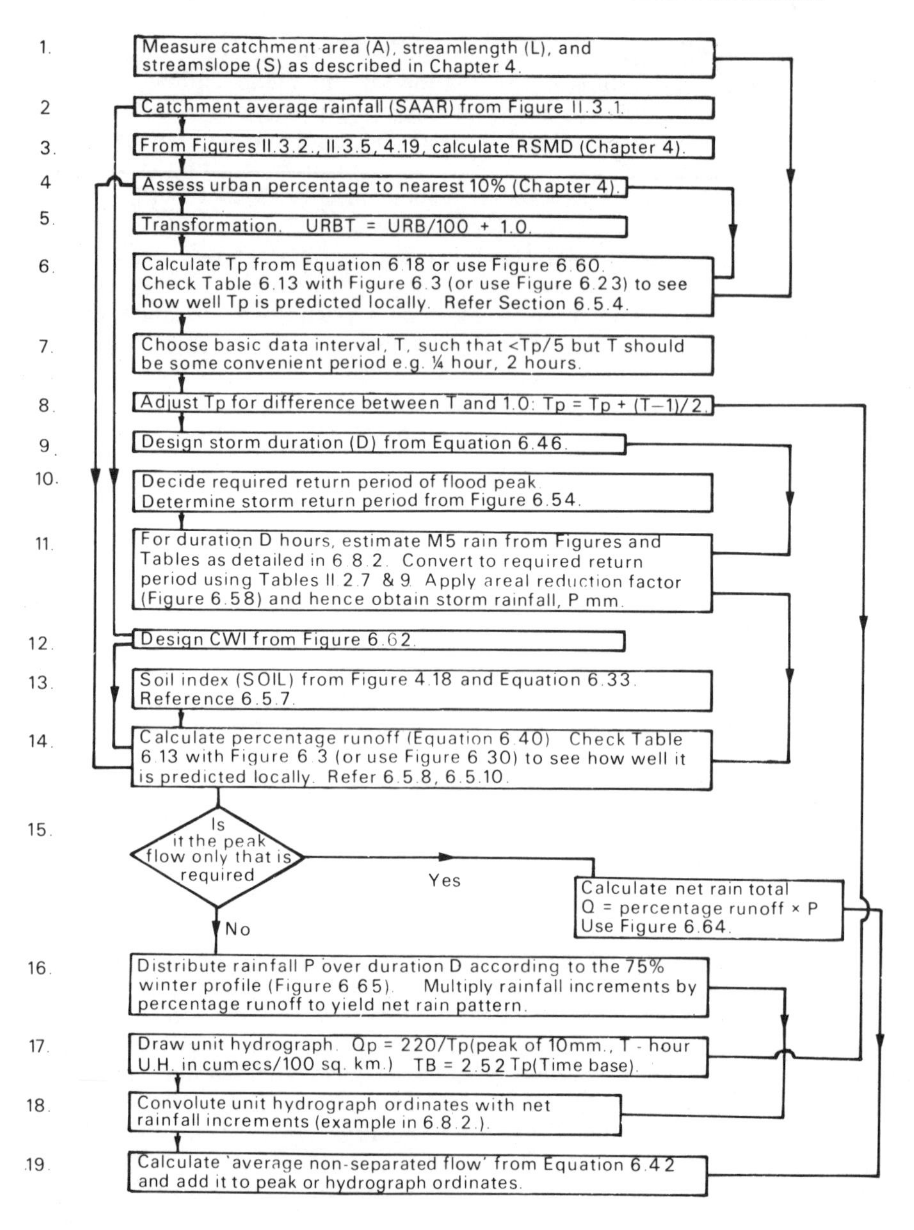

Figure 3.3 Flow chart of design procedures for flood with peak of specified return period (from Sutcliffe 1978).

Hydraulics Research Station (Price 1975, Price 1977a, Samuels & Price 1976). Systems for mathematically modelling river flows have been developed, many of which result from advances in computer use.

The model described by Samuels and Price (1976) is based on inputs in the form of observed or design rainfall. Hydrographs are then used to convolute the rainfall into runoff hydrographs for the main streams and the tributaries. Three computer programs are used – FLUCOMP1, FLUCOMP2 AND FLUCOMP3 – and Price (1977a) also describes a fourth, designated FLOUT (Price 1980). This last program is basically the river catchment model which combines the routing and surface runoff models developed for the *Flood Studies Report*. FLOUT is a more approximate method of routing than the FLUCOMP programs and can be used as an initial step in the FLUCOMP model or as the basis of elementary models (Hydraulics Research Station n.d.).

Price (1977b) has extended his work to consider the estimation of flood levels on floodplains from prescribed discharges. In this he uses three approaches, namely the use of historical records, the calculation of normal depths and the derivation of backwater profiles. This last technique is attracting interest among engineers despite the fact that many methods have been proposed over many years. The 'step' method (Chow 1959, Henderson 1966) is most commonly used and is particularly appropriate if there are likely to be errors in the normal depth calculation due to downstream effects (Price 1977b). Another contribution, concerning the relationship of inundating waters to patterns of floodplain topography, is that developed by Lewin and Hughes (1980). This comprises a qualitative inundation model involving types of input, output and transfer processes. The model has been applied to two rivers in Wales, the Dyfi and Teifi, although further research is required into applications to contrasting floodplain types (Lewin & Hughes 1980).

Another useful development since the completion of the *Flood Studies Report* is the work of Farquharson *et al.* (1975) who have derived a rapid prediction equation which may be of considerable value for practical application:

$$\mathrm{EMF} = 0.835\ \mathrm{AREA}^{0.878}\ \mathrm{RSMD}^{0.724}\ \mathrm{SOIL}^{0.533}\ (1 + \mathrm{URBAN})^{1.308}\ \mathrm{S1085}^{0.162}$$

where EMF is estimated maximum flow and the other parameters are as in the earlier equation (p. 64) and URBAN is the Urban Development Index (Sutcliffe 1978, p. 40). The equation has been derived from estimates of the maximum flow for 80 catchments by the unit hydrograph technique using observed values for the unit hydrograph and standard runoff estimates, which were then related to catchment characteristics to give a prediction formula (Sutcliffe 1978).

In addition to applications of the methods of the *Flood Studies Report* since its publication, and the experience gained from such applications (Marshall 1977, Welsh Water Authority 1977), further research and development of the methods used in the report have continued. A series of supplementary reports to the *Flood Studies Report* have been produced reflecting this further research. The use of the areal reduction factor, which has caused much discussion and indeed misunderstanding, has been further investigated (Institute of Hydrology 1977a) and an attempt is made to clarify the relationships between point and areal rainfall. Short return period floods have also been studied (Institute of Hydrology 1977b) and this is particularly important for the benefit assessments of flood alleviation schemes (Ch. 4), since the frequency of these floods generates a major contribution to any economic justification. Partial duration series methods are used for floods that occur with a frequency of greater than once a year (Parker *et al.* 1983a).

Several aspects of the *Flood Studies Report*'s regression equations have attracted attention, particularly the use of the regional multipliers. The dilemma of boundary catchments in the context of these multipliers, and also with respect to the regional growth curves, has been debated (Institute of Hydrology, 1977c). Catchment slope and area parameters outside the range of data used in the report have also been encountered and the problems this raises have been considered (Institute of Hydrology 1977c). Other supplementary reports relate to historical information for chalk catchments (Institute of Hydrology 1977d), design flood studies in catchments subject to urbanisation (Institute of Hydrology 1979) and in small catchments (Institute of Hydrology 1978a). Finally, research has developed a revision of the winter rain acceptance potential (Institute of Hydrology 1978b) and a comparison is presented between the *Flood Studies Report* unit hydrograph procedure and the rational formula (Institute of Hydrology 1978c).

A further review of progress is presented in *Flood Studies Report – five years on* (Institution of Civil Engineers 1981). Applications of *Flood Studies Report* techniques for specific aspects of engineering are discussed including reservoir safety (Law 1981), reservoir spillway design (Hallas 1981), hydroelectric schemes (Johnson *et al.* 1981) and reservoir operation for water supply and flood control (Mackey 1981).

Recent research advances relating to rainfall–runoff techniques are outlined by Lowing and Reed (1981), and statistical estimation is discussed by Beran (1981), particularly attempts to improve estimation of flood magnitude from ungauged catchments. Van Oosterom (1981) undertook a survey of Water Authorities and others who have used the report and discusses its uses and abuses. He indicates that the methods have been accepted in principle for control of floodplain development and have been used in some cases for Section 24(5) surveys, river improvement work and flood control schemes. Here, however, he reports that some Water Authorities, and especially those with long hydrological records, have continued to use their own methods. Van Oosterom (1981) also points to problems concerning catchment changes as a difficulty with statistical methods and to concern over the use of the methods for small and urban catchments. He suggests that the methods should be combined for the latter case with those outlined by Young and Prudhoe (1973) and Poots and Cochrane (1979). Finally, he reports that applying the methods to existing reservoir spillways constructed according to the Reservoir Safety Provisions Act (1933) is also problematic. Seven of the 20 spillways studies using the report's methods were found to be seriously deficient and, assuming a national survey produced similar proportions, the cost to enlarge the spillways to the report's recommendations would be very substantial (Van Oosterom 1981, Oates 1981).

THE RATIONAL FORMULA AND OTHER EMPIRICAL FORMULAE

An alternative method of flood analysis which has a long history is the empirical formula approach. The problems of these formulae are well known and largely concern the generalising of relationships, particularly between rainfall and runoff. Little account is taken of the basic hydrometeorological relationships involved. Francis (1973) indicates the problems clearly by applying different formulae to an imaginary circular 129 km^2 catchment in Britain. The results indicate, predictably, a wide range of values for the estimate of the maximum flood using the different methods. Nevertheless, these empirical methods are still used, particularly the 'rational formula', and support for their use has been expressed in various sources including the Proceedings of the Leningrad Symposium on *Floods and their computation* (IASH/UNESCO/WMO 1969) by Sokolov (1969) and Heras (1969).

Flood magnitude/catchment area relationships are some of the earliest approaches to flood analysis and a further development of such formulae incorporating an additional factor – namely rainfall intensity – is the 'rational formula'. This is sometimes referred to as the Lloyd–Davis (1906) formula and takes the form:

$$Q = C.\ i.\ A$$

where Q is the maximum flood discharge, C is the coefficient of runoff, i is rainfall intensity, and A is catchment area. The formula has been used for many studies, notably by the Institution of Civil Engineers (1933), and is still used, particularly for small catchments. The Institute of Hydrology points out that 'although the dimensional accuracy of the formula is well known, it is less well known that the formula can be regarded as the outcome of applying a rectangular unit hydrograph to a uniform rainfall' (Institute of Hydrology 1978c, p. 1). The conclusions from the comparison of the rational formula with the methods in the *Flood Studies Report* are that 'subject to an assumed use of identical runoff coefficients for small lowland catchments, the rational formula will yield peaks twice as large as those from the FSR' (Institute of Hydrology 1978c, p. 1). The major source of the difference is considered to be the use by most engineers of the Bransby–Williams formula for design rainfall duration in the rational formula (Bransby–Williams 1922). There has been no similar attempt to check or calibrate the rational formula despite its widespread use (Institute of Hydrology 1978c).

The US Department of Agriculture Soils Conservation Service (1957) developed a formula which incorporates catchment or contributory area, rainfall characteristics and a time parameter. Many other formulae take the form of relationships between catchment area, morphometric characteristics and an index of rainfall intensity and frequency (Potter 1961, Ward 1978). One such formula developed in Britain (Rodda 1969) uses a regression equation of the form:

$$Q = 1.08\ A^{0.77}\ R^{2.92}\ D^{0.81}$$

where Q is the mean annual flood, A is catchment area in square miles, R is the mean annual daily maximum rainfall and D is drainage density in miles per square mile. Rodda developed his analysis by relating the daily maximum rainfall for appropriate recurrence intervals to the T year flood and assuming equality between rainfall and discharge recurrence interval (Ward 1978). The result was further formulae:

$$Q_{10} = 1.22\ A^{0.69}\ R^{1.63}\ D^{1.06}$$

and

$$Q_{50} = 1.24\ A^{0.51}\ R^{2.02}\ D^{0.94}$$

Rostomov (1969) also produced a detailed empirical approach and, in summary, an equation of the form:

$$Q_{\mathrm{m}} = 16.67\ C\ B\ S\ A\ \frac{R_{\mathrm{A}}}{R_{\mathrm{d}}}$$

where Q_{m} is the maximum flood discharge, C is a runoff coefficient, B is a rainfall distribution coefficient, S is a catchment shape coefficient, A is catchment area, R_{A} is rainfall amount and R_{d} is rainfall duration.

Although empirical formulae, and particularly the rational formula, are likely to be used in the future, it seems clear that their use will give way to the techniques developed in the *Flood Studies Report* and in the Wallingford Procedure (see below). Given the results obtained in the assessment of the rational formula by the Institute of Hydrology (1978c) and Francis (1973) this is likely to lead to more satisfactory flood frequency and magnitude estimation.

URBAN STORM FLOOD ALLEVIATION: THE WALLINGFORD PROCEDURE

Dissatisfaction with the rational formula has led to revised methods for the analysis and design of urban storm drainage sewer systems (National Water Council 1981). The inadequate capacity of these systems can be a major cause of minor urban flooding, perhaps affecting just roads or a small number of properties, where sewer size is insufficient to discharge the sudden and intense runoff from heavy storms on an urbanised catchment.

The main contribution of the new Wallingford Procedure is a set of computer models to simulate the surcharge capacity of storm sewer pipes. This in turn is complemented by a further model to optimise sewer construction by examining the hydraulic efficiency of pipe systems with varying depth, gradient and pipe size. These varying characteristics lead to different costs which can be compared in the Wallingford Simulation Method with the benefits of storm sewer construction as measured by flood damages avoided.

Such trends towards full economic analysis of storm sewerage bring both technical difficulties and other problems. For example, the prediction of out-of-sewer flood volumes is possible with the Wallingford models but there is no systematic method of using these volumes to predict flood extents, which are needed for the analysis of flood alleviation benefits. Also it would appear that many storm sewer systems have been 'over-designed' in the past in that much has been invested to prevent little flood damage (Penning-Rowsell 1981b). Cost–benefit procedures have not traditionally been employed in this sphere and their introduction within the Wallingford system, logical though this is, will surely cause some soul-searching since they are likely to lead to lower design standards for urban storm sewers in order to reduce costs to match the relatively low level of benefits. Only time will tell whether such lower standards are acceptable to both professional engineers and the public.

Hydrographic analysis of coastal flooding

The techniques and studies discussed above have only concerned river flooding. However, in many parts of Britain, and particularly along the east coast of England, serious flood problems exist in coastal and estuarine areas. Flood information and techniques of analysis are just as important in this context.

Much historical evidence exists on east coast flooding although, as with other historical information, much of it is descriptive and imprecise. Nevertheless, Suthons (1963) obtained tide gauge levels with long records for a number of ports and he analysed the frequency of abnormally high sea levels on the east and south coast of England (with data from Kings Lynn, Harwich, Southend, Tilbury, North Woolwich, London Bridge, Chelsea Bridge, Sheerness, Newhaven, Portsmouth and Southampton). However, some of the longest records extending for up to 100 years have gaps and the longest continual set of records covers some 60 years up to 1962 (Suthons 1963). River flow records needed for the combined analysis of river flows and tidal surges are, of course, very much shorter. They generally date only from the 1950s and 1960s when river boards and river authorities started to develop their hydrometric networks.

The seriousness of the situation in the east coast of England was illustrated in 1953 when an event identified as a 300-year storm surge occurred (Departmental Committee on Coastal Flooding 1954). This caused the flooding of some 65 000 hectares along the coast from Tyneside to the Thames at a cost of approximately 300 lives. A popular account of the floods (Pollard 1978) points to the fact that in 1953 there was no mechanism whereby meteorological forecasts could be linked to hydrographic information to produce accurate predictions of coastal flood levels. The Storm Tide Warning Service was established after the 1953 floods, however, as an independent unit under the administration of the Ministry of Agriculture, Fisheries and Food Land Drainage Division. The service is based at the Meteorological Office for immediate access to up-to-date wind forecasts (Townsend 1980).

PREDICTING STORM SURGE LEVELS

The meteorological situation is critical for predicting the nature and effects of storm surges. These occur in the North Sea when low pressure systems move to the north of Scotland. Sea level rises from 305 mm for each 25 mm reduction in barometric pressure and this surge then runs down the North Sea producing a wave effect along the east coast of England which is accentuated as the depression passes (Undrell 1980). The situation is clearly most serious when the tidal surge coincides with high river flows in the Humber, Hull and other east coast rivers.

Two approaches can be used for forecasting tidal surges. The first involves empirical equations and this has been used for some time as the basis of the Storm Tide Warning Service forecasts for the east coast (Townsend 1980). The basis of present-day equations is that 'residuals' – the difference between actual tidal level and astronomically predicted tidal level – can be forecast for ports in the southern parts of the North Sea by using a combination of high water residuals at one or more ports further north and several wind parameters. Townsend illustrates this method with the equation for North Shields:

high water residual at Tyne = (0.75 × high water residual at Wick) + (0.03 × 320° component of surface wind at Fair Isle 11 hours earlier) + (0.002 × 030° component of geostrophic wind at 58°N 1°W 6 hours earlier) + 0.14 metres

Townsend (1980) points out that to use such equations the Storm Tide Warning Service has three main sources of information. Data are provided, firstly, by tide gauges at ten coastal stations, secondly, from tidal curves which provide hourly astronomically predicted tide levels for these ports and, lastly, from forecast winds from the meteorological forecasters. The use of empirical formulae in this context suffers from the same deficiencies as in other situations, namely that the constants chosen are empirically derived from data on past events and do not necessarily fit the whole range of conditions that can be met in practice. Townsend (1980) indicates that mean errors of 0.2 to 0.3 m occur and that very occasionally the error can be as much as 1 m. Naturally such large errors can critically affect the design of sea defence levels by requiring more freeboard time than would otherwise be the case.

The second approach to forecasting tidal surges uses recent developments in the field of mathematical modelling designed to produce a numerical solution using equations of motion and continuity describing the physics of the sea. The method involves a grid system covering the Continental Shelf around the British Isles and uses meteorological and sea forecasts in an iterative process. The process produces results in two forms: as 'residuals' at 30 ports and as contour charts of the developing surge (Flather 1980). Townsend (1980) reports that results with the model have been good in the first two years of operation. Rather better estimates have been obtained than when using the empirical formulae, despite some shortcomings in the model relating to difficulties of intervention with observed levels as the surge develops. Attempts have also been made to apply both an empirical formula and, more recently, a mathematical model to sea levels at west coast ports.

The statistical and theoretical problems of forecasting events combining both river floods and tidal surges are considerable. Working on the Yare basin in Norfolk, Mantz and Wakeling (1979) used two methods for estimating the design heights for 'safe' river bank crest levels. First, long-term annual maximum data and extreme value statistical methods were used to predict joint events of rainfall flood discharge and North Sea tidal surge at each end of the tidal catchment. Safe bank heights are determined for return periods of 25 and 100 years by producing a range of joint events which is applied to a finite difference mathematical model to find the maximum water levels at the return period. The second method, used to check the first, uses medium-term annual maximum data and extreme value analysis for eight stations in the Yare catchment study area.

FLOOD PROBLEMS AT KINGSTON-UPON-HULL AND LONDON

Of particular importance to east coast flooding is the evidence that the situation in the North Sea is not stable (Dunham & Gray 1972, Akeroyd 1972), and that the effects of tides and surges are becoming more pronounced. This situation is due to the 'secular rise' which has caused tide levels to increase along the east coast. This is generally attributed to the tilting of the land mass about an axis which passes approximately through Dundee and Anglesey, coupled with the partial melting of the polar ice caps (MacDonald & Partners 1978). As a result of rising tidal levels the traditional sea defences are slowly being rendered inadequate (Parker & Penning-Rowsell 1982) and estuarine barriers have been needed to prevent inland flooding.

One serious problem area occurs at Kingston-upon-Hull. The river Hull is tidal for some 30 km north of its junction with the Humber and surges of 1.8 m are not unknown. The situation is most serious when the tidal surges coincide with high river flows in the Humber and the Hull. The city of Kingston-upon-Hull lies at the mouth of the Hull on the north bank of the Humber and suffered serious flooding in 1969 (Fig. 3.4). A major flood-protection programme for Hull has been implemented which involves a barrier constructed by the Yorkshire Water Authority at a cost of £3.9 million and opened in 1980. This barrier comprises a vertical lifting gate 30 m wide, 10.6 m deep and 2.5 m thick, which can be shut in just 25 minutes to sustain a water loading of about 1200 tonnes (Freeman 1981). This design has retained the navigability of the river Hull and is coupled with the raising of sea defences along the banks of the Humber.

Other east coast barriers are being constructed or proposed but the most well-known for flood protection in Britain is the Thames flood protection barrier, designed to prevent the possibility of catastrophic flooding in London (Greater London Council 1970, 1971; Horner 1978, 1979). Horner (1978) has indicated that high tide levels at London Bridge have been rising (Fig. 3.5), reflecting both the secular rise of North Sea tide levels and the progressive constriction of the Thames estuary through the building of embankments.

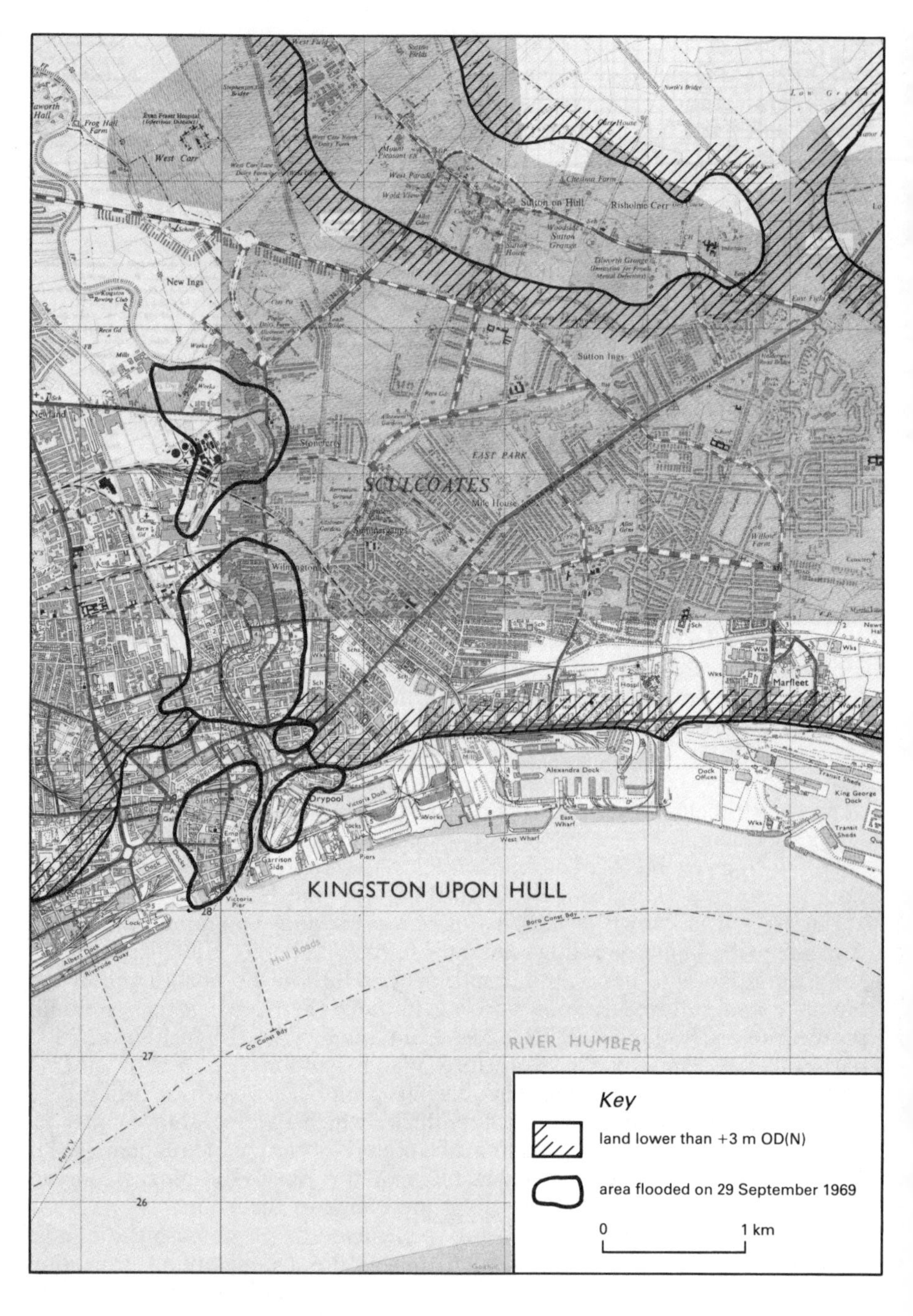

Figure 3.4 Flooding experienced at Kingston-upon-Hull in 1969 (from MacDonald & Partners 1978; Crown copyright reserved).

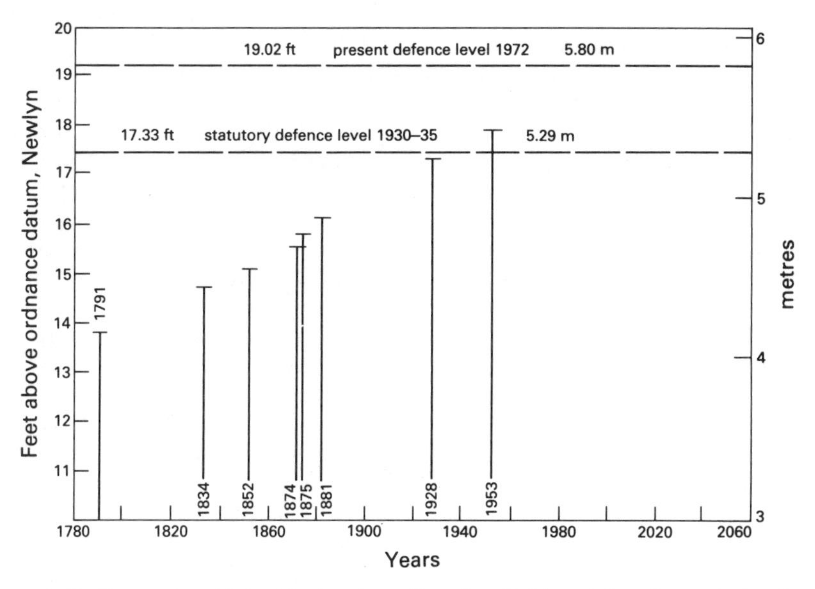

Figure 3.5 Increasing high tide levels at London Bridge (after Pratt & Holloway 1978).

Some 143 km^2 of London lies below high water level in the Thames Estuary, with a population of approximately 0.15 million, and tributary areas are also at risk (Butters & Lane 1975). Taken as a whole the area contains numerous important businesses, commercial and residential buildings which would be severely affected by a major flood (Bentham 1980). The Thames Barrier consists of a major scheme including a rising sector gate located at Silvertown in the western half of Woolwich Reach designed to allow navigation to pass normally. The gates are over 21 m high when raised but will only be raised from the river bed when storm surges are predicted. Extensive structural defences have also been provided for the downstream banks and the total cost of the scheme is estimated to be around £502 million at April 1981 prices. Of this sum £435.6 million is for the barrier and £62.5 million for the raising of the downstream defences. The remaining £3.9 million was used in the investigation stage and for interim bank-raising defence works.

Data on impeded soil drainage

The duration and frequency of soil waterlogging critically affects agricultural productivity (Trafford 1972, 1977). Agricultural drainage aims principally to reduce this waterlogging rather than overland flooding *per se*. In addition to their data on flood frequency and magnitude some information on soil waterlogging is contained in the Water Authorities' Section 24(5) surveys (see below) but this is mainly descriptive and rather uneven. The Nature Conservancy Council also undertakes wetland surveys to establish a scientific basis for a classification of wetland sites to be protected as nature reserves or scheduled as SSSI.

It is, however, mainly through the work of the soil survey that information on waterlogging has been improved. Until the mid-1960s most of their data related to visual observations of soils which were then placed into one of five classes of 'natural soil drainage'. The term was defined on the basis of depth to a grey horizon and the amount of organic matter in the upper layers. The visual technique was basically dependent upon the view that increased soil wetness was indicated by grey and ochrous colours at progressively shallower depths from the surface (Robson and Thomasson 1977).

In 1963 a change in measurement occurred to a method using more than 200 'dip-well' sites in Britain. From these and other sources more data have become available, mainly in the numerous 'soil survey memoirs'. In addition Robson and Thomasson (1977) have usefully brought together data relating to seasonal waterlogging in English lowland soils. The present handbook used by soil survey staff (Hodgson 1974) recommends using six classes of water regime with wetness duration measured at soil depths of 70 cm and 40 cm (Table 3.2). Robson and Thomasson (1977) define a 'wet soil' as containing water removable at a suction of less than 10 mb (10 cm water), a condition that is readily identified in the field.

Table 3.2 Soil moisture regime classes – wetness classes – duration of wet states (from Robson & Thomasson 1977).

Class	
I	the soil profile is not wet within 70 cm depth for more than 30 days* in most years†
II	the soil profile is wet within 70 cm depth for 30–90 days in most years
III	the soil profile is wet within 70 cm depth for 90–180 days in most years
IV	the soil profile is wet within 70 cm depth for more than 180 days, but not wet within 40 cm depth for more than 180 days in most years
V	the soil profile is wet within 40 cm depth for more than 180 days, and is usually wet within 70 cm depth for more than 335 days in most years
VI	the soil profile is wet within 40 cm depth for more than 335 days in most years

* The number of days specified is not necessarily a continuous period.
† 'In most years' is defined as more than 10 out of 20 years.

In the context of agricultural development soil survey waterlogging research has also aimed at determining a range of important physical soil properties including water retention capacity, porosity and the density of field soils (Hall *et al.* 1977) Much advice, of course, is also available to farmers on the drainage of waterlogged soils through the Agricultural Development and Advisory Service (1974–7) of the Ministry of Agriculture, Fisheries and Food. The series of 'drainage leaflets' explains all the major factors to be considered in assessing whether drainage works are feasible and desirable, including the economics of drainage, operations and techniques, and the availability of government grants. Technical advice is based on research into a wide range of problems relating to the drainage of agricultural soils undertaken at the Ministry of Agriculture's Field Drainage Experimental Unit (various dates) at Harpenden, Hertfordshire.

The interaction between agricultural drainage and flooding has been the subject of speculation for over a century and further research is still necessary. The controversy fundamentally concerns whether field drainage exacerbates flooding down stream through increasing runoff rates. Some research indicates that this is not the case (Green 1979), but that agricultural drainage reduces surface runoff and, although increasing the volume of total discharge, it flattens the flood peaks. Evidence also shows that for small fully drained catchments the height of flood peaks tends to rise (Green 1979) and this is confirmed for mole-drained plots (Robinson & Beven 1983), but uncertainty remains as to the effects of drainage on large catchments. For major events, when the catchments are in any case saturated, such drainage probably has little overall effect on flood volumes but may affect the timing of onset, as described for a small upland catchment by Robinson (1980). However, systematic data are virtually non-existent and the controversy remains. Nevertheless many land drainage engineers privately feel that there is an adverse effect on flood volumes from field drainage and from the reduction in floodplain storage with the protection of agricultural land from flooding.

Section 24(5) flooding and land drainage surveys

One of the major deficiencies affecting flood alleviation in Britain has been the sketchy and incomplete information available on flood hazard sites and land drainage problem areas. While many river authorities in the past have known in very general terms about their flood problems there has been until recently no consistency or indeed no precise detail in the information available. Nevertheless such material is essential and complementary to the hydrological information discussed above in providing data on the extent of flooding from which, amongst other things, an economic appraisal of the likely future flood damage or agricultural enhancement can proceed (Ch. 4).

THE TASK AND ITS CONTEXT

Section 24(5) of the Water Act 1973 requires Water Authorities 'to carry out, from time to time, and in any event at such times as the Minister may direct, surveys of their areas in relation to their land drainage functions' (Ministry of Agriculture, Fisheries and Food 1974). The surveys were designed to identify, map and tabulate data on areas that might benefit from flood protection or from agricultural drainage improvement to increase food productivity. Some commentators have seen the surveys as designed to perpetuate all aspects of land drainage work, and jobs, rather than for more altruistic motives.

Guidance notes issued by the Ministry of Agriculture, Fisheries and Food (1974) attempted to introduce a measure of consistency in the surveys between the various Water Authorities. However, the Ministry was perhaps not sufficiently aware of the constraints on Water Authorities in terms of

Table 3.3 Summary of Section 24(5) surveys of the Water Act 1973 (from Ministry of Agriculture, Fisheries and Food 1974, Penning-Rowsell & Chatterton 1976).

Mapped data (2/25 000)

1 main river
2 Internal Drainage Board areas and adopted watercourses
3 other problem watercourses
} Existing and proposed

4 irrigation channels, existing and proposed

5 pumping stations
6 sluices
7 weirs
} Existing and proposed

8 gauging stations
  (a) high flow
  (b) flood warning
} Existing and proposed

Overlays

1 areas liable to flood
  (a) normal floodplains
  (b) at risk from breached defences
  (c) at risk from subsidence
  (d) in risk of permanent inundation
  (e) liable to flood more than 1 m deep
2 areas where drainage is unsatisfactory (inadequate outfall)
3 Excessive surface runoff – existing and future
4 flood flow routes (with velocities, if they can be estimated)
5 flooding from culverts
6 affected property
7 duration of flooding (where this can be estimated)
8 floodplain zones, showing use (e.g. recreation, car park, agriculture) – existing and proposed
9 flood-proofed buildings – existing and proposed
10 proposed flood protection works

Tables (for each flood risk area)

1 nature of problem
2 population in areas liable to flood more than 1 m deep
3 risk to property, e.g. '6 houses', 'radio factory', 'agriculture', 'class B road'
4 estimated frequency (to approximate benefit/cost ratio and determine priorities)
5 estimated damage or improvement potential (£K)
6 warning systems – exists/proposed, with estimated cost (£K)
7 proposed works
  (a) frequency standard
  (b) type (widen and deepen; embank; flume; pump etc.)
  (c) estimated cost (£K)
8 proposed floodplain zoning
9 proposed floodproofing
10 estimated benefit/cost ratio
11 priority (1, 2 or 3)
12 proposed adoption as main river
13 Internal Drainage Board adjustment
14 bye-law adjustment

Appendices

1 probability – damage graphs (with bases)
2 sources of data, e.g. gauges, flood marks, newspaper reports, photographs
3 outline bases of costings
4 outline of bases of calculations of improvement to agricultural land

staff, time and money. Hence, the guidelines were over-ambitious (Table 3.3) (Penning-Rowsell & Chatterton 1976), although if all the reports had been produced in such detail a massive and immensely valuable source of information for flood hazard research and practice would have been available. In reality, although all Water Authorities have now produced Section 24(5) reports these vary in depth, scope and detail, and some are distinctly thin (Parker & Penning-Rowsell 1981a).

The reports were to have comprised, in great detail, maps of flood-prone areas and those liable to poor drainage, maps and details of main rivers, information on gauging stations, weirs, sluices and flood warning systems. Details were also to be provided on any proposed schemes for flood alleviation or drainage improvement. Overlays were to delimit floodplains and areas at risk from breached defences and give flood characteristics such as water velocities and flood durations, together with existing and proposed flood protection works. Tabulations were to detail each problem and alternative solutions with estimated costs and benefits. Finally, appendices were recommended containing sources of data, probability of damage and assumptions.

Water Authorities faced considerable problems in producing these reports. Few staff within some Authorities were trained in certain areas of expertise required for satisfactory completion of the task. Authorities themselves considered the guidelines to be over-ambitious and unnecessary. Particular areas which produced problems for some authorities were the generation of alternative plans for flood protection, including the less common non-structural adjustments, and the economic evaluation of the various schemes, particularly the assessment of benefits likely to accrue from different proposals. The cartographic expertise varies greatly between the Authorities and this is not an unimportant point when mapping clarity and accuracy is vital to the success of the programme.

Given the scope of the work it is perhaps surprising that the reports have been produced at all. It is clear, however, that some Water Authorities

Table 3.4 Section 24(5) reports: completion status and mapping scales (up-dated from Parker & Penning-Rowsell 1981a).

Water Authority	Completion status	Mapping scale
Northumbrian	B	Most scales from 1:1250 to 1:63 360
Yorkshire	A	1:50 000
North West	A	1:50 000
Severn Trent	A	1:25 000
Anglian	B	1:50 000
Welsh	B	1:250 000
Thames	B	1:10 560 and 1:25 000
Southern	B	1:25 000
Wessex	A	1:50 000
South West	B	1:100 000

A = completed and published in full according to Ministry requirements.
B = partly completed and published according to Ministry requirements.

undertook the exercise with little intention of attaining the goals set out in the Ministry guidelines, but aiming at more limited and less detailed surveys (Table 3.4). Nevertheless the survey results provide an invaluable data source on the major and the many minor flooding and land drainage problems in England and Wales.

EXAMPLES OF SECTION 24(5) SURVEYS

The Severn Trent Water Authority Section 24(5) report is one of the most comprehensive of such documents, consisting of some eight district volumes plus an atlas of maps with transparent overlays (Severn Trent Water Authority 1980). Detail on floodplain areas and sites protected from flooding are represented by overprinting. Other information presented includes Water Authority administrative areas, main rivers and the locations of weirs and interrogable gauging stations. Specific land drainage problem areas or 'benefit areas' are coded and mapped in the report, although flood characteristics such as depth and duration are not presented. The code-numbered land drainage problem sites are analysed in accompanying volumes containing descriptions and evaluations of each case. Where solutions to problems have been presented, the design standards and estimated costs and benefits are provided with benefit–cost ratio and a priority rating.

The Severn Trent report identifies National Nature Reserves and Sites of Special Scientific Interest, which were omitted from the original Ministry guidelines, and the possible effects on the nature conservation value of sites are included. An example of the detail provided is the case of benefit area 5 – 98 – 210 1 12 concerning river Derwent tributaries in Nottinghamshire. Because of urban floodplain encroachment and inadequate channels, parts of Long Eaton are flood prone; in a 100-year flood 1810 houses and 7 industrial premises would be flooded with estimated damages of £617 000 at December 1977 prices. The present value of the benefits of the recommended flood protection scheme is £864 000 giving a benefit–cost ratio of 1.4 : 1; work commenced on this scheme in 1980–81. On the other hand, a scheme for a nearby benefit area 6 – 98 – 810 – 2 only has a benefit–cost ratio of 0.1 : 1 and has little chance of being implemented.

At Shrewsbury on the river Severn (coded 1 83 – 510 – 16) there are up to 191 residential and 195 commercial properties which suffer from frequent flooding as also indicated by previous research (Harding & Parker 1974). This flooding begins when peak flows of 2½-year return period are experienced; flood depths greater than 1.5 m have been attained and the duration of flooding has exceeded 6 days with serious traffic dislocation. A major flood protection scheme is proposed in the Section 24(5) report consisting of channel structures at a cost of £1 615 000 (1977 prices). Discounted benefits are estimated at £4 220 000 giving a benefit–cost ratio of 2.6 : 1. The scheme is designed to contain at least the 100-year return flood period and the problem is coded 1A: top priority.

One of the more limited Section 24(5) reports is that produced by the Welsh Water Authority (1979). The analysis was constrained by the restricted staff resources the Authority was able to devote to the study and therefore the report falls short of the Ministry guideline requirements, with few detailed site assessments. The Authority admits that the survey does not fully cover Internal Drainage Board areas and that there has been little, if any, use of hydrological catchment studies to assess the effects of a design flood on new and existing developments. Sea defence studies have also been particularly limited despite a need for such surveys in the Authority's area. The Welsh Water Authority, however, underwent an internal reorganisation in 1980–81 and some of the deficiencies are being corrected.

Some of the most comprehensive Section 24(5) survey results come from Wessex Water Authority (1979). The example in Figure 3.6 shows the full range of agricultural and urban drainage problems. Some of these are quite minor but this is typical of the many small drainage problems within a large urban area: thus problem 163 is at a locality where just 27 houses are flooded in a 20-year event owing to a drainage channel with inadequate capacity to contain runoff. Problem 205, however, foresees a flood caused by a combination of high tides and storm runoff from the urbanised catchment whereby over 2000 properties in the central shopping and business area of Bristol might be directly affected. Property damage was originally estimated by Wessex Water Authority (1979) at £4.65 million from the 100-year event. Having identified this scale of problem the Authority then proceeded to more detailed feasibility studies for a comprehensive flood relief scheme, finding that property damage in the 100-year event is likely to be no more than £676 000 (Wessex Water Authority 1979, Parker 1983).

In terms of obtaining an overall picture of land-drainage problems one unfortunate aspect of the Section 24(5) surveys, given the Ministry guidelines, is the lack of consistency and the unstandardised nature of the surveys' mapping scales and detail. Only two Water Authorities, Severn Trent (1980) and the Southern Water Authority (1979), use the recommended basic mapping scale of 1 : 25 000, although much of the Thames area is also mapped at this scale (Thames Water Authority 1978). The Welsh Water Authority locate their problem sites on 1 : 250 000 sheets, although they hold unpublished material mapped at a larger scale, and the Northumbrian Water Authority (1978) use maps of various scales.

A further matter for concern in terms of flood research is that data on flood depths, duration of flooding and flood flow routes vary greatly between Authorities. Clearly the end product of the survey programme is something well short of a systematic hazard zone atlas which could have resulted (Parker & Penning-Rowsell 1981a). Furthermore, different Authorities identify problems differently. In the Wessex Water Authority (1979) survey the river Avon floodplain near Fordingbridge in Hampshire is mapped at 1 : 100 000 scale. Numbered triangle symbols indicate flooding and agricultural drainage problems and, where they are large enough, the SSSIs are indicated. No attempt is made, however, to quantify the severity of impeded soil drainage, perhaps as the average number of weeks of soil saturation, so that comparison could be made with agricultural drainage problems in other parts of the country. Consistency in the analysis of such problems could have helped greatly in setting priorities for future expenditure, and it is unfortunate that the attempts by the Ministry of

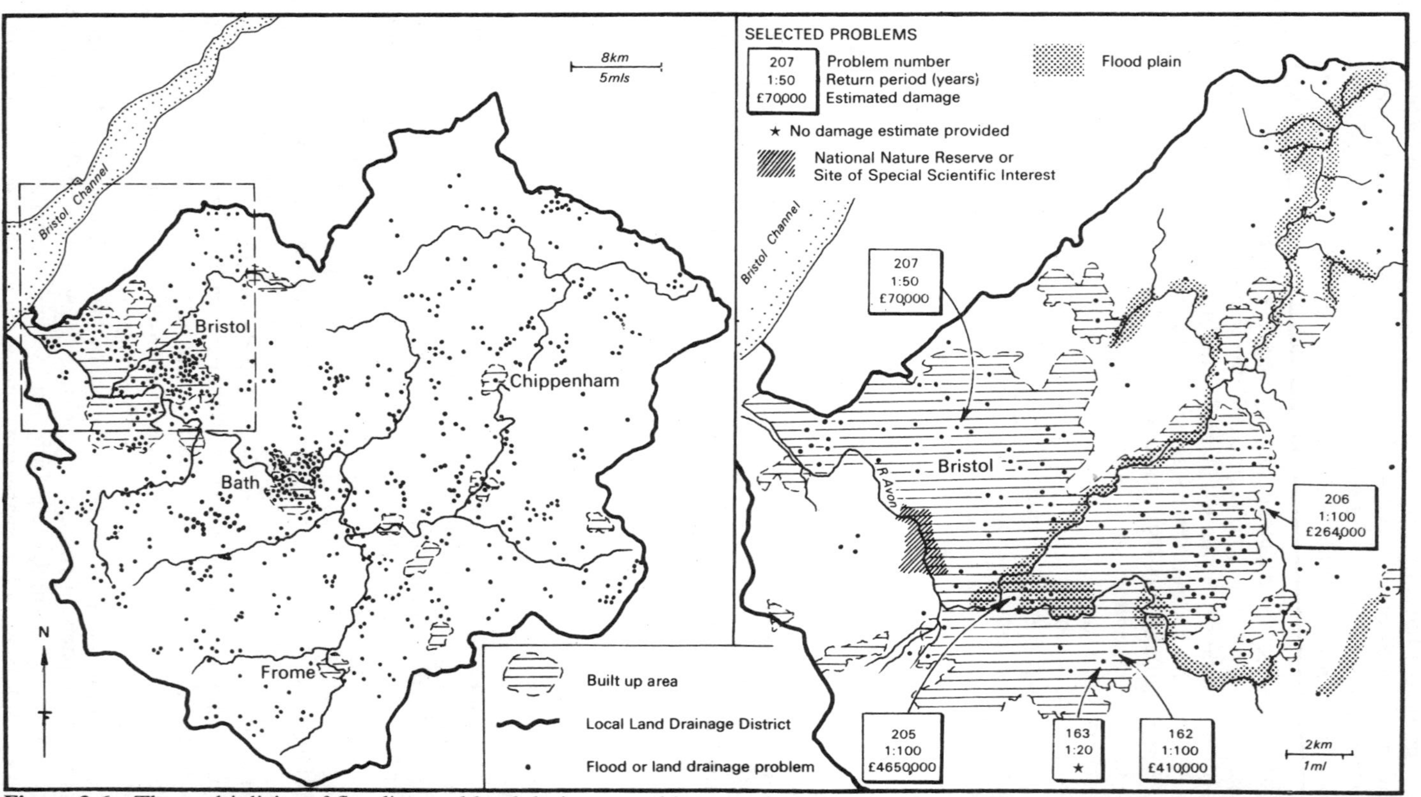

Figure 3.6 The multiplicity of flooding and land drainage problems as shown on the Wessex Water Authority (1979) Section 24(5) surveys.

Agriculture, Fisheries and Food to set universal standards did not meet with greater success.

Assessment

There can be no doubting the very real advances made in Britain within the last decade in the fields of hydrometry and data-collection, research and practical application of methods for flood prediction and estimation. This work has all improved the information base on which flood alleviation and agricultural drainage programmes can be founded.

The data gathering programmes of various kinds are continuing to yield information and as records get longer more detailed analysis will be possible which in turn will lead to more accurate flood predictions. The increase in the density of hydrometric networks in the 1970s has slowed, partly because Authorities consider their networks are reasonably complete (Van Oosterom 1981), but also because financial constraints in some cases prohibit further instrumentation. Gaps that still exist are likely to continue and therefore data deficiencies will always occur in studies of particular localities. A remaining problem relevant to flood alleviation and floodplain development concerns the translation of flood discharges at a site into flood extents. This is not covered by the *Flood Studies Report*. Although research is developing at the Hydraulics Research Station (Price & Samuels 1980, Samuels 1981, Samuels & Price 1981), no readily available method exists as yet for this important procedure.

Section 24(5) survey reports are, to some extent, already outdated as a result of floods that have occurred since the data collection was undertaken. In these surveys Water Authorities largely defined problems in terms of the maximum floods for which records exist and the occurrence of floods of longer return periods will necessitate updating. In particular, data on long return period floods are limited, partly because historical records for many sites are limited or unreliable; more attention should also be directed to lesser floods as these critically affect the worthwhileness of flood alleviation schemes. No systematic attempt has as yet been undertaken to simulate the effects of severe design storms and thereby predict the 'maximum probable flood' as is common in the United States (United States Army Corps of Engineers 1978) where the folly of basing floodproofing and land-use regulations on the 100-year flood extent has been recognised following floods exceeding this frequency.

To be questioned, however, is the extent to which an improved data base and improved techniques of flood analysis and estimation will, in practice, improve planning and management. There tends to be a belief among many scientists and decision makers that once data and information are available the planning and management process will somehow automatically be improved. This is not necessarily so. There is a clear need to incorporate the data and information obtained from surveys and monitoring into a thorough analysis and discussion of priorities, design standards and alternative strategies for both flood alleviation and improved agricultural productivity. Only in this way will the data obtained serve effective planning and floodplain management, rather than dominate it.

4 *The economics of flood alleviation and land drainage*

The need for economic evaluation

Governments and water planning agencies in most countries recognise the need for some form of economic evaluation before deciding to construct a flood-alleviation scheme or to devise some form of non-structural alternative (e.g. Ministère de l'Equipment et du Logement 1969, Ministry of Agriculture, Fisheries and Food 1974, United States Water Resources Council 1979, Office of Public Works (Eire) 1974, Department of National Development and Energy (Australia) 1981). There is increasing awareness in Britain and elsewhere of the need to determine the economic effectiveness of agricultural improvements such as land drainage as part of a wider process of public sector policy evaluation (Ministry of Agriculture, Fisheries and Food 1974, 1978; H.M. Treasury 1982). Nevertheless the limitations of such economic evaluation are increasingly recognised (Self 1970, Edwards 1977, Sandbach 1980, O'Riordan & Sewell 1982). We must therefore proceed with great care, recognising both the conceptual and technical problems involved in valuing environmental goods and the effect that cost–benefit analysis can have in making what are essentially political decisions appear comfortably neutral, scientific and therefore unquestionable.

The use of economic evaluation stems fundamentally from the need to gauge the returns the community or individuals receive from flood alleviation and land drainage in relation to the costs incurred. This may simply mean assessing whether the benefits obtained exceed the costs, or gauging the likely 'loss' to be incurred if it is decided to proceed for 'political' reasons with some form of scheme despite the costs exceeding the benefits. Systematic economic evaluation is particularly necessary in this field of environmental planning and public investment since it is difficult to make sound intuitive judgements about the worthwhileness of flood alleviation schemes given that such judgements involve the assessment of very low probabilities. Formal methods of evaluation at least standardise information gathering and presentation before political considerations are involved in decision making.

A more sophisticated aim of economic evaluation than simply ensuring that the benefits exceed the costs involves selecting the level of community investment in flood alleviation and agricultural drainage which shows the greatest economic return. This occurs when successive increases in the standard of the scheme or non-structural alternative no longer yield proportionate benefits. Such evaluation requires a full analysis of the changing costs and benefits with different scheme standards and as yet this practice is in its infancy (Local Government Operational Research Unit 1973, 1978, National Water Council 1981).

Economic perspectives

THEORY

Cost–benefit analysis conventionally aims above all else at maximising the economic efficiency of investments. Such analysis is a tool for simplifying decision making by producing simple indices of economic worthwhileness such as the benefit : cost ratio. It is not a means of conflict resolution. Cost–benefit analysis is a branch of applied welfare economics and is concerned therefore with the welfare of society. This welfare must be measured, somehow, from the preferences of individuals. Given the judgement in neo-classical economics that the individual is the best judge of his or her own welfare, these preferences are taken to reflect the 'utility' of the choices on offer.

Many conceptual and technical problems exist in cost–benefit analysis. The first is how to aggregate individual utilities or satisfactions, in our case in the form of flood damage avoided, so that the economic effects on all those affected can be counted together. The problem would be avoided if the marginal utility of money income were the same for everyone: if an extra pound of flood damage avoided were worth the same for a pauper as for a millionaire. If a simple addition of utilities were then accepted, the total welfare is simply the addition of money values across all beneficiaries or contributories to a flood alleviation or drainage scheme. This assumes, however, that the distribution of money income is already optimal in maximising social welfare, otherwise a redistribution of incomes would also result in increased social welfare. If income distribution is not considered optimal then adjustments to cost–benefit analysis or a different evaluation procedure are required. Otherwise projects will be evaluated according to the existing sub-optimal income distribution, and therefore the inappropriate utilities as far as maximising social welfare is concerned.

A further set of problems concerns which distribution of costs and benefits to aim at, and the related conflicts between economic efficiency and equity. A 'pareto improvement' occurs where someone gains from investment but no-one loses and this appears uncontroversial (Thompson *et al.* 1983). Investment should increase until there are no more such gains without losses: the point of the 'pareto optimum'. However, this criterion becomes almost impossible to meet in practice. Most projects of any size will adversely affect a minority, for example when a riparian owner has to sacrifice part of a garden to build a flood containment wall. Nevertheless this criterion does clarify the mind, puts flesh on the question 'Welfare for whom?' raised in Chapter 1, and initiates a search for those disadvantaged in the interests of some 'common good'.

The criteria actually used in cost–benefit analysis, however, is the Hicks–Caldor compensation principle. A project passes the test if the gainers could compensate the losers and still be better off, but actual compensation is not paid (Pearce 1977). Mishan (1971) notes that a potential pareto improvement could involve the rich getting richer at the expense of the poor but insists that cost–benefit analysis carries no distributional significance. It only shows 'that the total of gains exceeds the total of losses, no more' (Mishan 1971, p. 404).

Pearce and Nash (1981) argue, however, that Mishan's approach to judging economic efficiency depends on an assumption that the existing income distribution is optimal, and that such assumptions reflect only one of many distributional judgements, each of which is arbitrary. Some form of weighting system of costs and benefits is therefore required to correct for a sub-optimal income distribution. This would entail, for example, weighting more highly the flood damage incurred by the poor than that by the rich. The basis for this is the likelihood that the effect, or disutility, of a given amount of flood damage for the former may be devastating, but for the latter it could be insignificant. Weights could be derived from marginal tax rates, although other factors affect these rates rather than just equity considerations. Furthermore, cost–benefit analysis might be used to evaluate schemes that deliberately set out to redistribute income and therefore change marginal tax rates. Weights can also be derived from past government decisions although this assumes that past judgements will continue to be valid for the future and implies that these decisions took income distribution into account and were equitable (in which case why bother with cost–benefit analysis?).

The approach advocated by Sugden and Williams (1978) appears to be a middle course between Mishan (1971) and Pearce and Nash (1981). It sees cost–benefit analysis as not purely concerned with economic efficiency but embraces distributional judgements, depending upon the decision makers' objectives. An unweighted potential pareto improvement might be the appropriate criterion used by those making individual investment decisions, were the government, at a strategic level, to use fiscal policies to convert this potential improvement into an actual improvement, perhaps by levying a betterment tax on those gainers and thus paying compensation to the losers. Within Sugden and Williams' scheme cost–benefit analysis plays an active rôle in the decision makers' choice by clarifying the distributional implications of these decisions. The objectives are at the decision makers' discretion, but are made explicit. Once these objectives and weighting system have been determined the analysis ensures consistency and accountability. As a consequence cost–benefit analysis as a technique loses its appearance of impartiality, but in any case this is largely false.

ECONOMIC AND FINANCIAL APPRAISAL

A fundamental difference exists between financial appraisal and economic analysis (Local Government Operational Research Unit 1978). The former examines decisions from the viewpoint of an individual or single enterprise and judges whether flood control or other investment is worthwhile to them. The latter attempts to examine the returns to the community at large resulting from community investment.

The two approaches can give very different results. For example, an agricultural drainage scheme may well be worthwhile for a small group of farmers when their costs of field drainage are far outweighed by the benefits they receive as the market value of increased output. However, the scheme may well not be economically efficient. The community pays the farmers' grant on their field underdrainage works and all of the arterial costs. In addition, a part of the market price the farmer receives is a direct subsidy from the government. In these circumstances the private gain to the farmer as measured in financial appraisal can far overestimate the community's return from the investment decision.

Taking another example, a flood alleviation scheme for just one factory might be financially viable from the industrialist's perspective: the benefits are obtained at very little or no cost. From the community or taxpayers' perspective, the scheme may not be economically worthwhile because the disruption caused by flood damage to the factory's production is not a real loss since this production may be made up elsewhere in the country.

When evaluating flood alleviation or land drainage schemes for government grant decisions, or from other community perspectives, economic analysis should be used, not financial appraisal. However, with economic analysis it is by no means clear that market mechanisms comprising many private transactions can adequately measure community costs and benefits (Edwards 1977). Thus the market value of a flood damaged carpet may be very low. To an impoverished householder the effect of the flooding may be negligible, in that unfortunately they may be accustomed to dirty surroundings, yet the cost of replacement might be a major proportion of their wealth. To a millionaire the dirt and damage might be intolerable yet the cost of such a purchase might be trivial, so what is the real value of the carpet and its flood damage?

ACCOUNTING PRINCIPLES

Cost–benefit analysis is not accounting, but economic perspectives lead to certain accounting rules.

The first of these is the 'with and without' criterion (Regan & Weitzell 1947, Eckstein 1968, Howe 1971, Mishan 1971, Pearce 1976, 1978). The analysis for investment decisions should concern itself solely with the difference between the situation *with* the investment and that *without* it. No costs and benefits should be included unless the costs are genuinely incurred in alleviating the flooding or improving agricultural drainage and the benefits are resources saved or created with the scheme or non-structural alternative.

Thus, for example, the cost of amenity tree planting associated with a flood alleviation scheme is not expenditure necessary for flood alleviation. The costs of field rationalisation in conjunction with agricultural drainage are not necessarily essential to realising the benefits of that drainage. These investments should not therefore be counted as scheme costs to be compared with scheme benefits, but should be the subject of separate appraisal. Similarly the costs of floodplain zoning or emergency relief should include only the necessary marginal increase in local authority administration overhead costs, not their total amount.

When counting the benefits of flood alleviation it is incorrect to take full replacement cost of damaged household items (Penning-Rowsell & Chatterton 1977). These items have given service for the period until they were damaged by flooding. The resource cost to the community if the items are totally damaged is not their full replacement value but their depreciated value, even if to replace them costs the individual flood victim considerably more. Similarly, the benefits of agricultural drainage are not the full value of the increased output net of subsidies, but the increase in value over and above the generally increasing agricultural production even without drainage.

A second set of accounting rules require appropriate comparison: we must compare like with like. Thus, for example, expenditure at current prices should not be compared without adjustment with benefits occurring some time in the future. Both must be reduced to their present values or some other reference point in time using techniques such as discounting. Transfers between different parts of the economy, including taxes and subsidies, should not be confused with real resource gains and should be excluded from both costs and benefits. As we have seen it is also incorrect to compare costs incurred by the community (social costs) with larger gains realised by just one section of that community (private benefits) and necessarily to conclude that the scheme is economically worthwhile.

These accounting 'rules' reinforce the need to be aware that parameters within an economic evaluation are not absolute but vary for different times, different individuals, different communities and different geographical scales of analysis. Indeed it is one of the major tasks within economic evaluation to standardise these parameters as far as possible so as to minimise the distortions that otherwise arise.

Economic evaluation of urban flood alleviation schemes

THE 'IDEAL' FRAMEWORK

The economic evaluation of urban flood alleviation schemes involves a number of practical difficulties but the theoretical framework is quite clear (Kates 1965, James & Lee 1971, Local Government Operational Research Unit 1973, Penning-Rowsell & Chatterton 1977, 1980, Cole & Penning-Rowsell 1981). The term 'urban' here embraces all dwellings, shops and factories where flood damage may occur, including those in rural areas.

The three data components essential to the evaluation (Fig. 4.1) comprise flood probability data (Fig. 4.1B), often derived from river flow or hydrographic records (Ch. 3), data on flood stage for given probabilities (Fig. 4.1A) and flood damage data for different flood stages (Fig. 4.1C). These data combine to yield a notional figure for the annual average flood damage to be avoided – or the annual scheme benefit – calculated as the area under the loss-probability graph in Figure 4.1D.

The annual benefit figure averages the damages from all floods of all frequencies. The benefits of a particular flood alleviation scheme comprise the difference, up to its design standard level of protection, between the annual average damages without the scheme – or without the non-structural alternative – and the annual average damages after flood alleviation is implemented. Allowance for flood damages after implementation is necessary because no structural flood-alleviation scheme can prevent all flood damages, and some indeed may cause massive but infrequent damage if overtopped (Chatterton & Farrell 1977). There are therefore inevitably

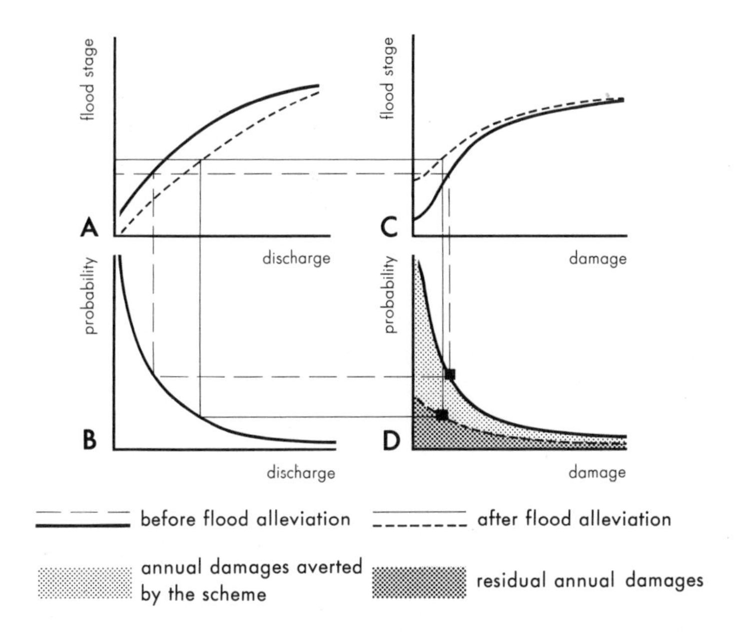

Figure 4.1 Theoretical relationships between flood probability, flood damage and the benefits of flood alleviation.

residual potential flood damages (Cole & Penning-Rowsell 1981) and these cannot be included as scheme benefits since they are not to be prevented.

The framework encapsulated in Figure 4.1 does not, however, lead to automatic decision making; there may well be a number of schemes where the benefits exceed the costs and a decision has to be made which to implement. This decision can be assisted, in theory, by examining the marginal increase of benefits with increasing costs as reflected in higher standards, as shown in Figure 4.2. For all levels or standards of protection between *X* and *Y* the benefits exceed costs. Within that range, however, the best value for money if budgets are limited would be scheme 'S' where the greatest benefit is obtained at least cost and the benefit–cost ratio is at a maximum. If finances were unlimited, however, then scheme 'T' should be chosen since here the difference between costs and benefits, or the net present value, is greatest: the 'profit' on the investment is at a maximum although the total investment cost is higher.

EVALUATION TECHNIQUES IN PRACTICE: DATA PROBLEMS

The theoretical niceties of Figures 4.1 and 4.2 are impressive. Seldom, however, can such analysis be complete and therefore the full aims of economic evaluation can rarely be met. A major problem remains the lack, or inaccuracy, of the necessary data.

For information on flood probability and flood stage we generally have to rely on historical data from gauged streams or other records to predict the

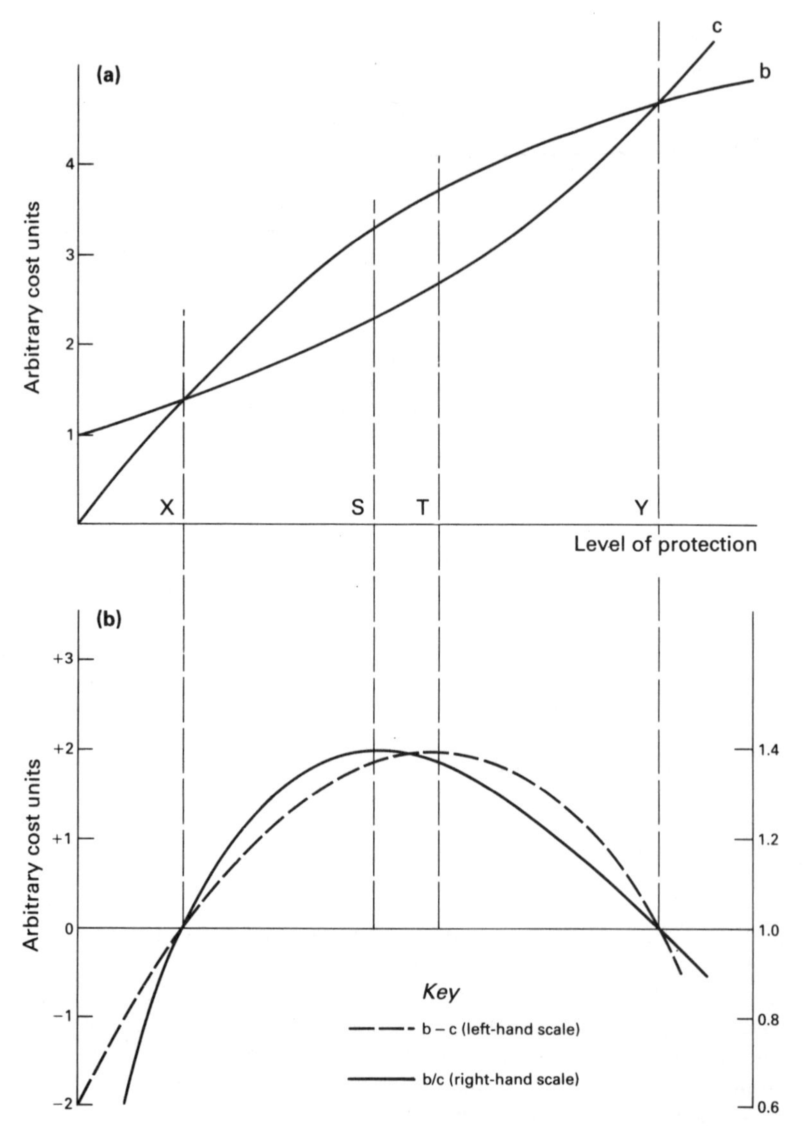

Figure 4.2 Theoretical relationships between the benefits and costs of flood alleviation to varying standards (after Local Government Operational Research Unit 1973).

extents of flood waters for future events of different return periods (Ch. 3). Given these data, flood depths within properties on the floodplain can be calculated, assuming a reasonably uniform flood profile, and damage estimates made from depth/damage data for individual properties (see Fig. 4.3).

However, historical records of floods with long return periods are sparse (Potter 1978). The more frequent events will occur with somewhat random flood extents, as local hydraulic anomalies within the floodplain create different flood patterns with each event. Moreover, flood studies have traditionally collected data on flood discharge rather than extent (Ch. 3). The floodplain may also have been affected by urban encroachment or perhaps partial flood control schemes. Recourse to generalised simulations such as of the maximum probable event is common in the USA (United States Corps of Engineers 1978) but this is difficult in Britain given the small and densely occupied floodplains. Physical or mathematical models through which floods can be routed are expensive. All this means that predicting future flood outlines within which to estimate future damages is not easy.

Calculating the benefits of flood alleviation requires data on future flood damages to be averted. Pioneer work in the USA developed the use of depth/damage functions which recognise that damage increases when properties are flooded to greater depths (White 1945, Kates 1962, Smith *et al.* 1979, Smith 1981, Parker & Penning-Rowsell 1982). This technique has been developed in Britain by Porter (1970) and Parker (1976) who measured actual flood damage in relation to flood depth.

However, the use of actual flood damage data derived from past events has problems. Unless carefully researched much flood damage can pass undiscovered or improperly counted, especially damage to the fabric of buildings which may only appear months after a flood. Damage assessments immediately after flood events can exaggerate losses prior to a careful analysis of salvage values. These problems have led to the development of 'synthetic' standard depth/damage relationships. These are based on a careful synthesis of actual damage information and published data on household inventories or retail stock levels which gives 'standard' data with wide applicability (Penning-Rowsell & Chatterton 1977, 1980) (Fig. 4.3). These data are linked to the land use of areas affected, recognising that different land uses or property types have different levels of susceptibility to flood damage (Penning-Rowsell 1976).

Nevertheless there is a danger in using standard flood damage data unquestioningly and ignoring the important conceptual problems, assumptions and technicalities inherent in their compilation (Parker & Penning-Rowsell 1983, O'Riordan 1980c). For example, the standard dwelling or building types used in the research may not be fully appropriate to the local circumstances in which potential flood damage data are required. Some properties, owing to the large quantity and high quality of their contents, may have their damage potential underestimated when using the standard data. Conversely, dwellings with unusually few or poor quality contents may have their flood damage potential exaggerated.

A related problem concerns predicting the effect of flood warnings. Warnings will reduce damage and recent research has quantified this reduction (Penning-Rowsell *et al.* 1978, Cole & Penning-Rowsell 1981). Thirteen locations throughout England and Wales were studied and 160 people were interviewed who had experienced flooding. Of those receiving a flood warning some 46 per cent did not react with damage-reducing actions: either they were too old or infirm, or the flood took them by surprise despite the warning. Many were sceptical of the warning given, having received false warnings in the past. Many taking emergency action to reduce damage

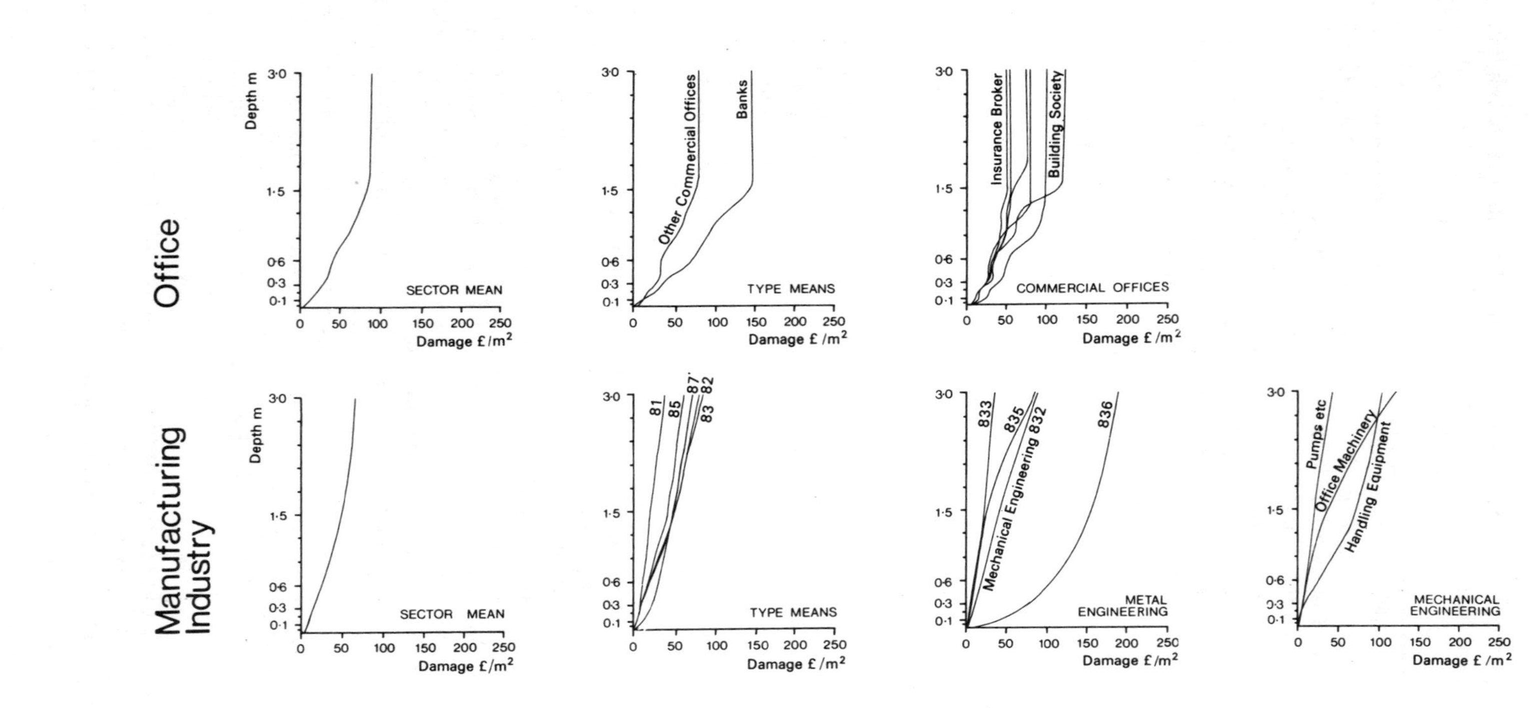

Office
Manufacturing Industry
Depth m
3·0
1·5
0·6
0·3
0·1
0
50
100
150
200
250
Damage £/m2
SECTOR MEAN
TYPE MEANS
Other Commercial Offices
Banks
COMMERCIAL OFFICES
Insurance Broker
Building Society
81
85
87
83
82
METAL ENGINEERING
833
835
Mechanical Engineering 832
836
MECHANICAL ENGINEERING
Pumps etc
Office Machinery
Handling Equipment

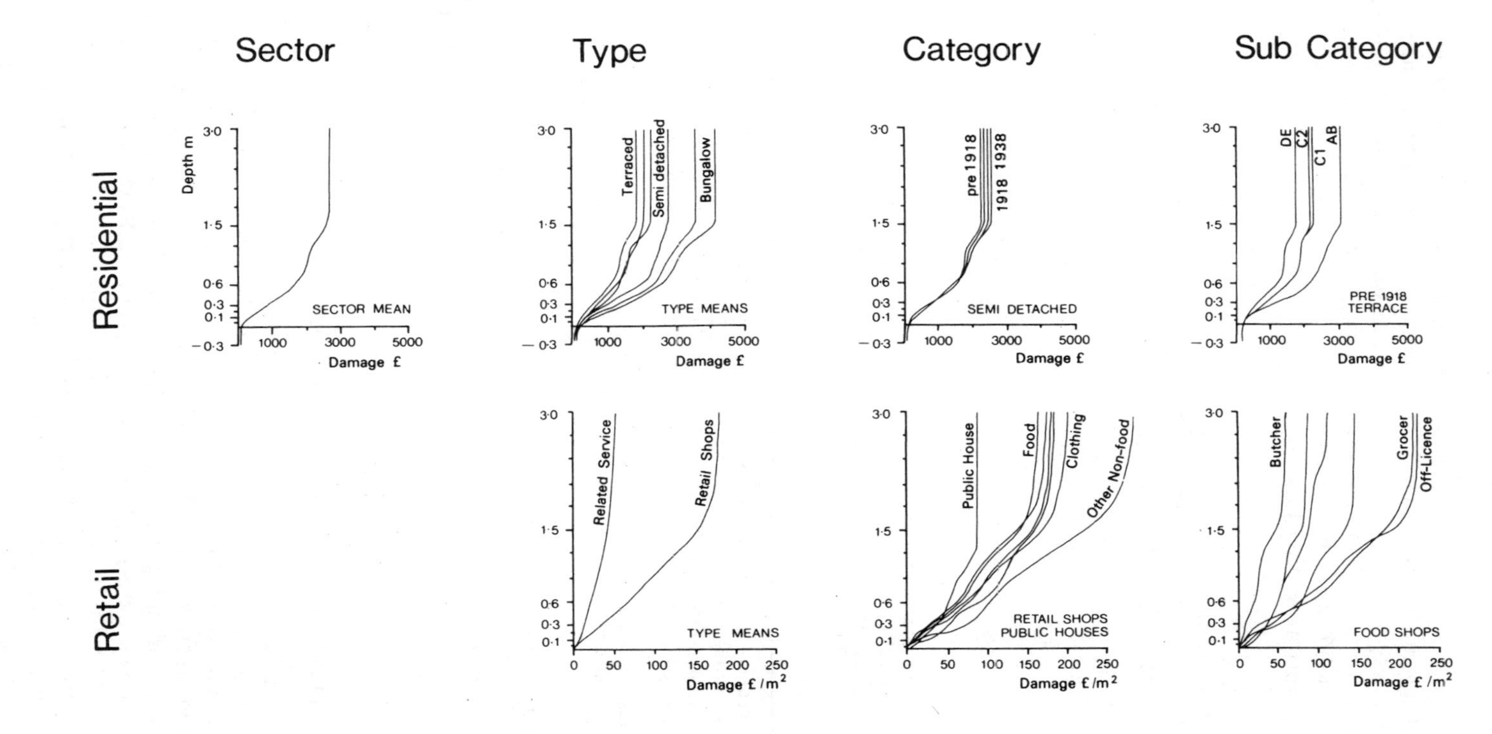

Figure 4.3 Depth/damage data for selected properties in Britain (from Penning-Rowsell & Chatterton 1977; 1977 prices).

obtained little benefit, for example because household effects were still damaged despite being moved. The relatively low average damage saving from these warnings (Table 4.1) counters to some extent the contention by Smith and Tobin (1979) that flood warning systems are a substitute for structural flood alleviation schemes, but of course such systems are invaluable for preventing loss of life.

Table 4.1 Generalised data on the benefits of flood warnings measured as damage reduced (January 1977) (from Cole & Penning-Rowsell 1981).

Residential property						Retail shops				
		Estimated average damage reduction with flood warnings[†]					Estimated average damage reduction with flooding warnings[§]			
Depth of flooding (m)	Total potential damage*	Up to 2 h warning		2–4 h warning		Total potential damage[‡]	Up to 2 h warning		2–4 h warning	
		£	%	£	%		£	%	£	%
1.2[¶]	2220	400	18	550	25	7999	1700	21	2400	30
0.9[¶]	2030	400	20	550	27	6327	1300	21	1800	28
0.6[¶]	1740	300	17	450	26	4604	900	20	1200	26
0.3[‖]	908	450	50	500	55	2318	1000	43	1100	47
0.1[‖]	338	150	44	150	44	424	150	35	150	35

* From Penning-Rowsell and Chatterton (1977), Appendix 2.3.
† Damage saving rounded to nearest £50.
‡ From Penning-Rowsell and Chatterton (1977), Appendix 3.1.
§ Damage saving rounded to nearest £100 for depths 0.3–1.2 m, to nearest £50 for depth 0.1 m.
¶ 70% response rate estimated from survey results.
‖ 70% response rate and 33% successful flood-proofing estimated from survey results.

Nevertheless the effectiveness of flood warnings is highly variable because the message dissemination from the forecaster to the public is often poor. The responsible water authorities may commit substantial investment to their flood forecasting capability yet the dissemination 'path' is neglected to the detriment of the overall warning efficiency and hence damage reduction (Penning-Rowsell *et al.* 1983).

THE COMPLEXITY OF INDIRECT FLOOD DAMAGES

In certain circumstances indirect damage from floods can be a considerable addition to direct damages (Green *et al.* 1983b, Parker *et al.* 1986). Indirect damages comprise traffic disruption, loss of retail and industrial output and the costs of emergency operations (Table 4.2). The use of simple fixed factors to relate direct and indirect damages is inadvisable (Butters & Tuck 1973) since indirect damages are highly site specific (Penning-Rowsell & Parker 1980, Green *et al.* 1983b). One assessment of flooding in the central area of Bristol has thus related indirect effects, measured as disruption time, to estimated direct damage to give the following formula:

$$Y = 0.808 + 0.000029K + 0.0000133F \ (r = 0.91; n = 33)$$

where Y is the period of industrial disruption (weeks), K is the direct damage to capital goods and F the direct damage to finished goods (£), each of which is significant at the 0.01 level or better (Parker *et al.* 1983a).

Table 4.2 A classification of indirect effects of flooding

	Likely approximate resource effect	
	Resources required	Resources lost
(a) Disruption of industrial and commercial business		
industrial production (manufacturing)		⊛
industrial production (service sector)		⊛
retail/commercial turnover		⊛
(b) Disruption of communications		
road traffic (including postal services, public transport etc.)	*	?
rail traffic	*	?
(c) Disruption of public utilities, etc.		
sewerage and sewage treatment	+	?
gas provision	+	?
electricity provision	+	?
telecommunications	+	?
(d) Cost of emergency operations, etc.[1]		
water authority	**	
hospitals	**	
police	**	
fire service	**	
ambulance service	**	
miscellaneous voluntary services	**	
military personnel	**	

[1] Some or part of these might also be considered to comprise direct damage (e.g. repair of damaged flood defences).
⊛ Lost profit or equivalent.
* Fuel for diversions etc.
+ Extra maintenance costs.
? Revenue loss.
** Staff overtime and other marginal costs.

Given that indirect flood damages result from this disruption of economic activities, two possible adjustments occur within the economy to minimise losses. Firstly, the disrupted activity may be deferred in time so that, for example, the retail trade lost by a flooded shop may take place after the flood has subsided and therefore no real loss results from the disruption. Secondly, the activity may be transferred elsewhere: industrial production or road traffic affected by a flood may simply be moved to a flood-free location. There may be costs incurred as a result of this transfer, perhaps in the form of increased travel distances, but a particular factory's turnover lost during a flood may well exaggerate the benefits of flood alleviation.

Viewed in this way indirect flood damages are a function of market failure. Economic theory would predict that the disruption to industrial production caused by flooding would immediately and automatically be compensated for elsewhere as firms there, competing for the markets, seize the opportunity to increase their production. In reality, markets do not adjust so rapidly and therefore there are real economic losses through disruption rather than complete transfers of economic activity during floods.

Vulnerability to indirect flood losses may be generalised, however, as an equation as follows (Green *et al.* 1983b):

$$\text{vulnerability} = f\,(\text{dependence, transferability, susceptibility})$$

Dependence measures the degree to which an economic activity requires a particular commodity as an input, such as a continual supply of electricity or labour, which might be disrupted during a flood. Transferability is the ability to transfer the activity elsewhere or to defer it in time. Susceptibility measures whether a flood will have an impact on the economic activity in question.

To illustrate this, when flooding occurs at Portland a large naval base is affected (Penning-Rowsell & Parker 1980). This base is *dependent* for its labour force on road communications with the mainland since the causeway is very *susceptible* to flooding. However, the Navy operates a boat service for key workers during floods. This *transfers* the base's reliance on the causeway to another transport medium thus reducing, but not eliminating, its vulnerability to indirect flood loss.

COMPUTER ANALYSIS

The calculation of annual average benefits is complex and a computer 'model' has been developed to simplify these economic evaluations (Chatterton & Penning-Rowsell 1978, 1981). This model incorporates land use data for the floodplain area, depth/damage data for relevant properties and adds indirect damage data to calculate annual average damages and discounted benefits (Fig. 4.4). Many options are available to suit local hydrological data, damage characteristics, flood warning times and alternative scheme life and discount rates (Figs 4.4 & 7).

A major benefit of computerised economic evaluation is that it allows for easy sensitivity analysis of the effect on calculated benefits and costs of key assumptions in their compilation. In addition it allows easy identification of particular beneficiaries of flood-alleviation schemes and thus the distributional impacts of the possible investment. This can be used to seek equity of impact, however defined, or to seek contributions to the costs of a scheme in proportion to the benefits those affected will obtain, in line with the fourth principle underlying land drainage law (Ch. 2).

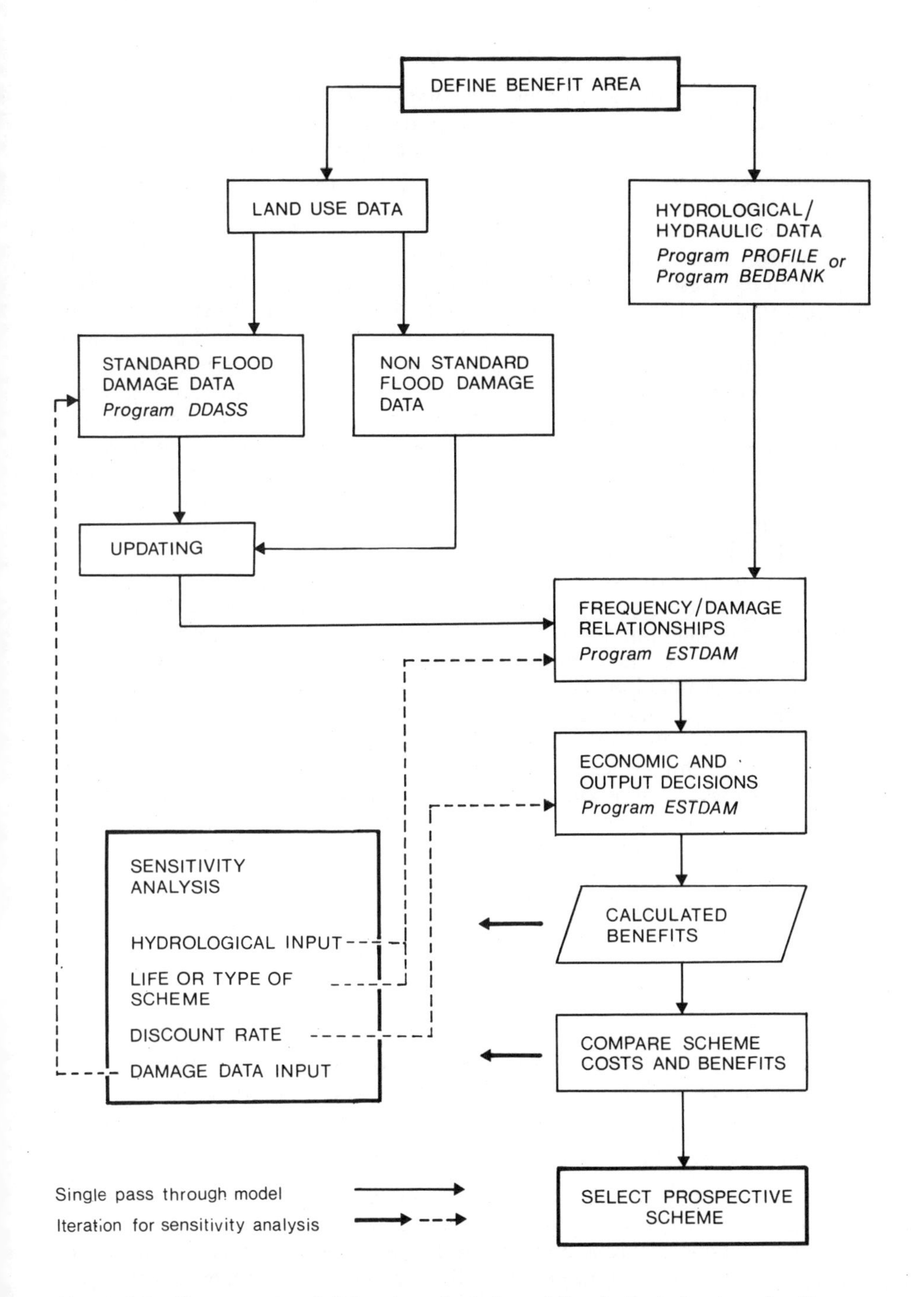

Figure 4.4 Computer 'model' for the calculation of flood-alleviation benefits (from Chatterton & Penning-Rowsell 1981).

DISCOUNTING AND EVALUATIVE INDICES

The final process in the economic evaluation of flood alleviation benefits is discounting the benefits over the anticipated lifetime of the scheme. Here it is assumed that the annual average flood damages will occur each year during the scheme's life. No other assumption is possible since future floods are not predictable.

Discounting, however, is widely misunderstood. The aim is to compare the worth of flood alleviation with other investments (Local Government Operational Research Unit 1973, H.M. Treasury 1982). If flood alleviation were forgone, then the resources saved could be used elsewhere, perhaps for industrial investment or for hospitals. Such investments would show a return and it is argued that flood alleviation should compete for scarce capital resources and pay a price for using its share; this price is expressed by the discount rate as a rate of return. Thus, if the return from industrial investment were 5 per cent in real terms, ignoring inflation, then someone investing £100 now would receive £105 after one year. Therefore having £100 now or £105 in one year's time are equivalent. The discount factor or rate is used to reduce future sums, whether costs or benefits, to their equivalent present values (i.e. the £105 to the £100).

Alternatively expressed, a discount rate higher than zero expresses the public rate of time preference; the rate at which the public values having money for consumption now rather than in the future. A high discount rate means money in the future is worth much less than money today, irrespective of inflation (H.M. Treasury 1982). As such it discourages long-term investment and encourages current consumption. Rates of discount appropriate to both flood alleviation and agricultural drainage are somewhat controversial and critically affect calculated costs and benefits (Cole & Penning-Rowsell 1981). However, the Treasury insists on a constant rate for all public sector projects – currently 5 per cent – despite this rate being higher than the average rate of return on private sector investment (Penning-Rowsell & Parker 1983, Fig. 7.2).

After discounting, various indices can be calculated to summarise the economic evaluation (Mishan 1971) (see Figs 4.2 & 11). A value for the benefit–cost ratio exceeding 1.0 indicates that the investment is economically worthwhile, other things being equal. The net present value is the difference between the costs and the benefits and shows the 'profit' or 'loss' the scheme would involve. The internal rate of return expresses the annual benefits in relation to the capital costs and shows the rate of return on the investment for comparison with alternative uses of the same financial resources.

THE PROBLEM OF INTANGIBLES

Many of the benefits of flood alleviation comprise the reduction of the intangible effects of flooding (Green *et al.* 1983a), including worry about impending floods (threat anxiety) and stress during floods events (event anxiety) (Table 4.3). Some floods in Britain are severe enough to cause loss of life, such as at Lynmouth in 1952, on the east coast in 1953 and in London in 1927. During the confusion that frequently surrounds flood

Table 4.3 The social benefits of flood alleviation.

1 avoidance of loss of life during flooding
2 avoidance of physical injury during flooding
3 alleviation of fear, anxiety and concern regarding flooding
4 avoidance of ill-health either due to flood risk or following flooding:
(a) psychiatric problems: tension; breakdown, depression;
(b) physical illness: ailments and death
5 avoidance of inconvenience which flooding causes: the 'destruction' or loss of time
6 avoidance of loss of community (extreme cases only)
7 avoidance of deprivation of quality of life experiences (e.g. holidays missed)
8 avoidance of losses in the value of historic buildings and conservation areas
9 avoidance of other miscellaneous impacts: unpredictable site-specific factors; loss of memorabilia
10 enhanced amenity

events people may act 'irrationally' and this may endanger their lives. For example, during the storms and flooding on the Isle of Portland in 1978 a man drove his car into Portland Harbour and drowned in the confusion of a damaging storm.

The most thorough research in Britain concerning the health effects of floods is reported by Bennet (1970) who compared the health record of those affected by the 1968 floods in Bristol with a control group outside the flooded area. Interviews with flood victims showed a significant increase in physical ill-health among males and psychiatric ill health among females. A significant increase in surgery attendances by males was noted from general practitioner records. Although women affected by the flooding attended more surgeries than those unaffected, the difference between flood victims and the control group was not statistically significant. The numbers of hospital referrals, however, more than doubled in the year after the floods for the flooded group, with again men showing the largest increase. Bennet (1970) also found a 50 per cent increase in mortality in the year after the flood for those affected, with deaths rising from 58 to 87 persons. The most pronounced rise was in the males of age group 45 to 64 for whom deaths rose nearly threefold from 7 to 20.

Similarly thorough research is presented for Australia (Smith *et al.* 1980, Handmer & Smith 1983), where hospital records were examined for Lismore, New South Wales, following the serious flooding in 1974. Here the floods appeared not to have affected the number of hospital admissions or deaths, but the pattern of admissions altered. From the most severely flooded houses the number of males admitted to hospital doubled, but the number of female admissions halved. However, the use of just hospital admissions is a severe test of the health effects of floods. Abrahams *et al.* (1976) found that the number of visits to general practitioners, hospitals and specialists all significantly increased for those affected following the 1974 flood in Brisbane. Smith *et al.* (1979) found that some 5 per cent of interviewees in the flooded area of Lismore reported 'serious illness' resulting from the flooding and 8 per cent reported some adverse health effects. These effects were more marked amongst the elderly and those most severely flooded.

Attempts to obtain self-assessment values for the inconvenience of floods (Sterland 1973) have yielded estimates for comparable flood situations as divergent as £200 and £10 000 (Penning-Rowsell & Chatterton 1977, p. 116). More recent research has related a variety of intangible effects of coastal flooding at Swalecliffe, Kent, to respondents' rating of the event's overall impact (Parker *et al.* 1983b). Over 50% of households said they still had not recovered from a flood 5 years earlier. Overall, the relative severity of the impacts, from greatest to least, was as follows: disruption of the flood; loss of memorabilia (small sample); leaving home; stress from the flood; worry about future flooding; damage to contents; health effects; and damage to the house. Those reporting the most overall impact were also those reporting the most serious health effects, stress from the event itself and having to leave home (Table 4.4). What is notable about these results is the relative unimportance of direct flood damage or, conversely, the greater importance of intangibles within the totality of flood impacts.

These findings do not negate the economic evaluation of flood-alleviation schemes but point out that damage measurable with current techniques is only one of many flood effects. Measurement improvements can be made: the valuation of life is commonplace for life assurance companies and the courts frequently value illness or disabilities when awarding compensation. It is clear nevertheless that gauging the cost-effectiveness of flood alleviation measures is not a precise science. This lack of precision hinders meeting the objective of determining the most worthwhile level of expenditure, as

Table 4.4 Correlations between householders' assessments of the overall impact of flooding and the different consequences (from Parker *et al.* 1983b).

Correlation of overall severity with:	Correlation coefficient (Pearson's r)	F ratio and significance level	Sample size*
damage to house	0.40	7.5 (< 0.01)	42
damage to replaceable furniture and contents	0.56	18.6 (< 0.01)	43
loss of irreplaceable objects	0.33	0.8	9
health affects	0.69	38.1 (< 0.01)	44
stress of flood	0.72	45.8 (< 0.01)	45
having to leave home	0.69	16.5 (< 0.01)	20
disruption	0.61	23.8 (< 0.01)	43
worry	0.76	58.7 (< 0.01)	45
other	0†		

* Number of households who rated this consequence other than 'not applicable'.
† Only four households reported and specified an additional type of loss for 'other'.

opposed merely to judging whether predetermined scheme designs show benefit–cost ratios exceeding 1.0. It also reinforces the need for thorough debate concerning the appropriateness of flood alleviation rather than mechanistic reliance upon calculated indices.

EVALUATION TECHNIQUES IN PRACTICE: KEY LESSONS

During research at Middlesex Polytechnic several lessons have emerged concerning the practicalities of the economic evaluation of flood alleviation schemes. The first of these involves the loss-probability relationship and particularly significant here are the return period and damage estimates for minor floods. Because of their frequency these contribute substantially to the annual average damages yet these floods tend to be least well recorded and ignored by the professional engineer. Moreover the shape of the curve often has to be approximated. The curve in Figure 4.1D is the ideal but that in Figure 4.6 is more usual. Owing to insufficient flood extent and probability data the 'curve' is made up from only a few points joined somewhat arbitrarily with straight lines. The calculation of annual damages is therefore less accurate than if fuller information were available.

Secondly, the availability of better potential direct flood damage data in Penning-Rowsell and Chatterton (1977) has produced a tendency to ignore the intangible flood effects. Previously these were accounted for in Britain by crudely doubling the estimated direct flood damages. Since 1977 the decision makers and the Ministry of Agriculture, Fisheries and Food have perhaps too readily thought that the published 'standard' flood damage data encapsulates all the possible effects of floods.

Thirdly, design engineers tend to aim at flood alleviation standards that are too high. Politicians are also particularly prone to advocate complete protection from flooding. Although design standards appear to have risen through time (Penning-Rowsell 1982a) the schemes that are designed to high standards – perhaps the 100-year flood event – can be economically unjustifiable. Considerable waste of time and resources can occur in attempting their justification. A related lesson is that feasibility studies of flood alleviation schemes should evaluate a range of design standards rather than just a single arbitrary design flood. Part of this process can be to undertake the economic analysis at different levels of detail (Penning-Rowsell & Chatterton 1980) rather than committing substantial resources to detailed economic analysis before it can be seen to be worthwhile. The analysis of benefits and the engineering design process can then proceed iteratively, to optimise design against the background of the economic results, rather than economic evaluation being used to seek justification for predetermined designs, as is all too common today.

Two case examples: Lincoln and Chiswell, Dorset

LINCOLN: DISTRIBUTIONAL EFFECTS

The only recorded major flood on the river Witham through Lincoln was in 1947 (Fig. 4.5). However, upstream river improvements giving flood

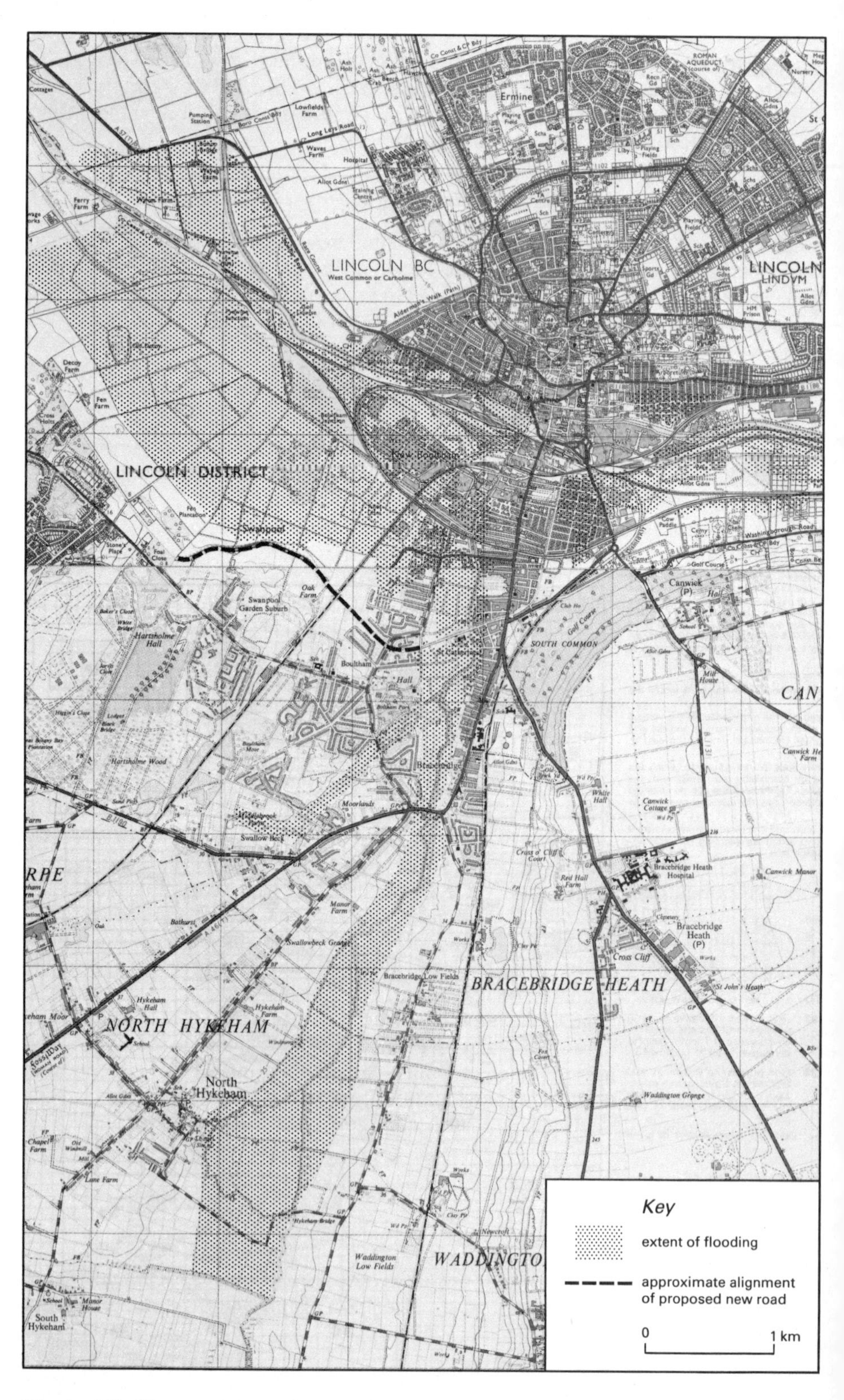

Figure 4.5 Extent of flooding in the 1947 flood at Lincoln (from Anglian Water Authority 1982; Crown copyright reserved).

protection to agricultural land appear to mean that the historic pattern of floods spreading on to rural washlands has been replaced by the bulk of floodwaters passing downstream to affect Lincoln. Here the floodplain is encroached by 19th century low-income housing concentrated around large factory complexes providing the town's main employment.

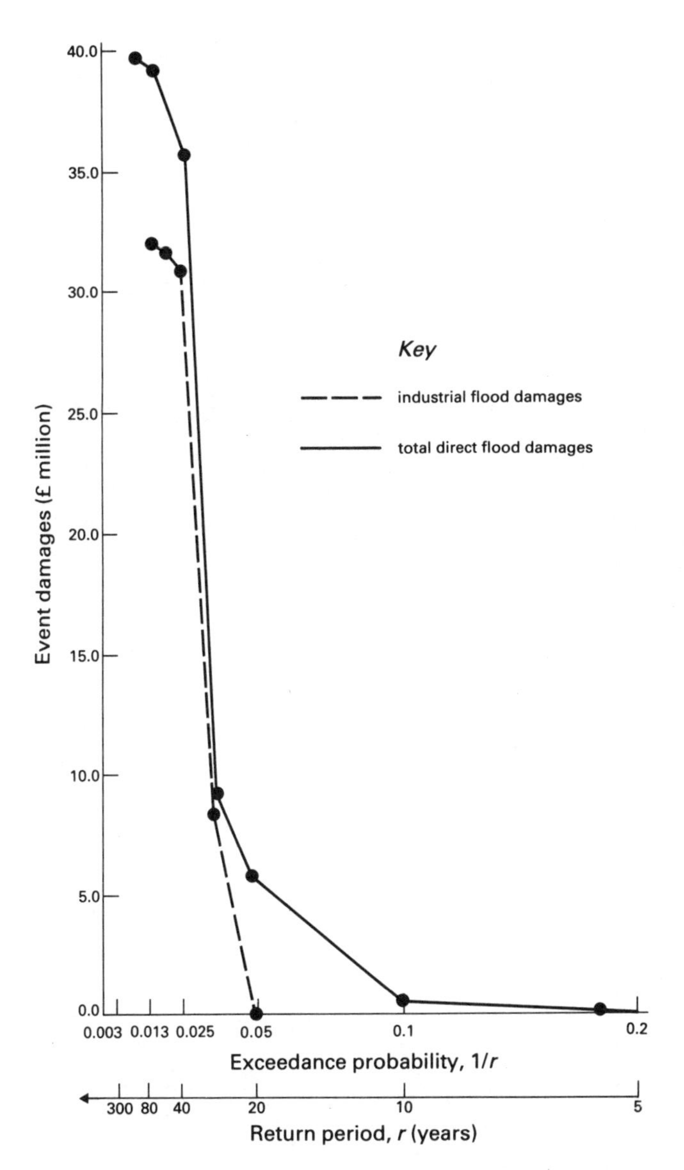

Figure 4.6 Lincoln flood alleviation scheme: loss–probability 'curve' (direct damages only: see also Fig. 4.7).

RETURN PERIOD YEARS	EXCEEDANCE PROBABILITY	BENEFIT (£)	PROBABILITY OF FLOOD IN INTERVAL	AVERAGE BENEFIT (£)	INTERVAL BENEFIT (ANNUAL) (£)	INTERVAL BENEFIT (DISCOUNTED) (£)	CUMULATIVE BENEFIT (ANNUAL) (£)	CUMULATIVE BENEFIT (DISCOUNTED) (£)
5	0.20000	0						
			0.10000	249777	24977	495741	24977	495741
10	0.10000	499554						
			0.05000	548770	27438	544586	52415	1040327
20	0.05000	597987						
			0.01667	4271571	71192	1413012	123607	2453339
30	0.03333	7945155						
			0.00833	22009132	183409	3640285	307016	6093624
40	0.02500	36073109						
			0.01071	37726324	404210	8022723	711226	14116347
70	0.01429	39379540						
			0.00429	39589133	169667	3367535	880893	17483882
100	0.01000	39798726						

Figure 4.7 Lincoln flood alleviation scheme: output from computer model assessing the benefits of flood alleviation (showing benefits of £17 483 882 deriving from event damages of £39 798 726, both for the 100-year design standard event).

Feasibility studies by the Anglian Water Authority (1982, 1983) used a mathematical model to simulate future flooding using minor events in 1980 and 1981 for model calibration. Predictions of flooding for a range of return periods were made for 31 'cells' or compartments of the floodplain. Flood levels in these cells were used to compute potential flood damages using the Middlesex Polytechnic assessment model, which indicated for the 100-year event that some 3000 houses, 151 shops and 44 factory units would be affected. The total predicted event damages for that flood would be some £32 million at 1981 prices and discounted annual average benefits totalled approximately £18.6 million (Figs 4.6 & 4.7).

However, these 'gross' benefit figures conceal a feature of major distributional significance. Some 85% of total industrial potential flood damage, or 70% of total potential damage, occurs to one firm, Ruston Bucyrus, which is one of the world's largest manufacturers of excavators and cranes. Technically this property could be given individual flood alleviation and leave the houses unprotected. This 'divisive' solution was not adopted, however, despite it almost certainly being the most economically efficient. Instead the Water Authority chose to subsidise the protection of the residential property from the benefits of protecting the industrial premises. The decision has also effectively subsidised from community funds – the land drainage precepts and government grant aid – the removal of the private industrialist's flood problem.

The situation was complicated in 1984 when the farmers owning the land designated for upstream flood storage, including the Lord Lieutenant of Lincolnshire, forced a public inquiry in an attempt to increase the compensation they had been offered. By this time, moreover, some of the industrial premises justifying the scheme had closed, reflecting the national and world-wide economic recession. The economic and political viability of the Anglian Water Authority scheme was thus threatened by the twin pressures of potentially increasing financial costs (of compensation) and reducing benefits, the latter brought about by the structural change of the British economy away from manufacturing. This change in turn was probably forced or accelerated by sterling's 'petro-currency' status and the 1980's trend towards balance of payments surpluses created by North Sea oil exports. In the end, however, the scheme proved economically viable because the stability of construction costs during Britain's recession off-set the fall in benefits from the redundant factories. The farmers 'won' the contest by negotiating a near doubling of the compensation on offer.

THE CHESIL SEA DEFENCE SCHEME: ASSESSMENT OF INDIRECT FLOOD EFFECTS

Flooding from the sea occurs regularly in Chiswell, Dorset, due to *percolation* through Chesil Beach and occasional overtopping of the Beach by *storm surge* waves. Very rarely *ocean swell* flooding occurs when sea water crashes over the Beach causing considerable damage and major erosion of beach material (Penning-Rowsell & Parker 1980, Lewis 1979) (Fig. 4.8).

These three types of flood were the only ones available on which to base

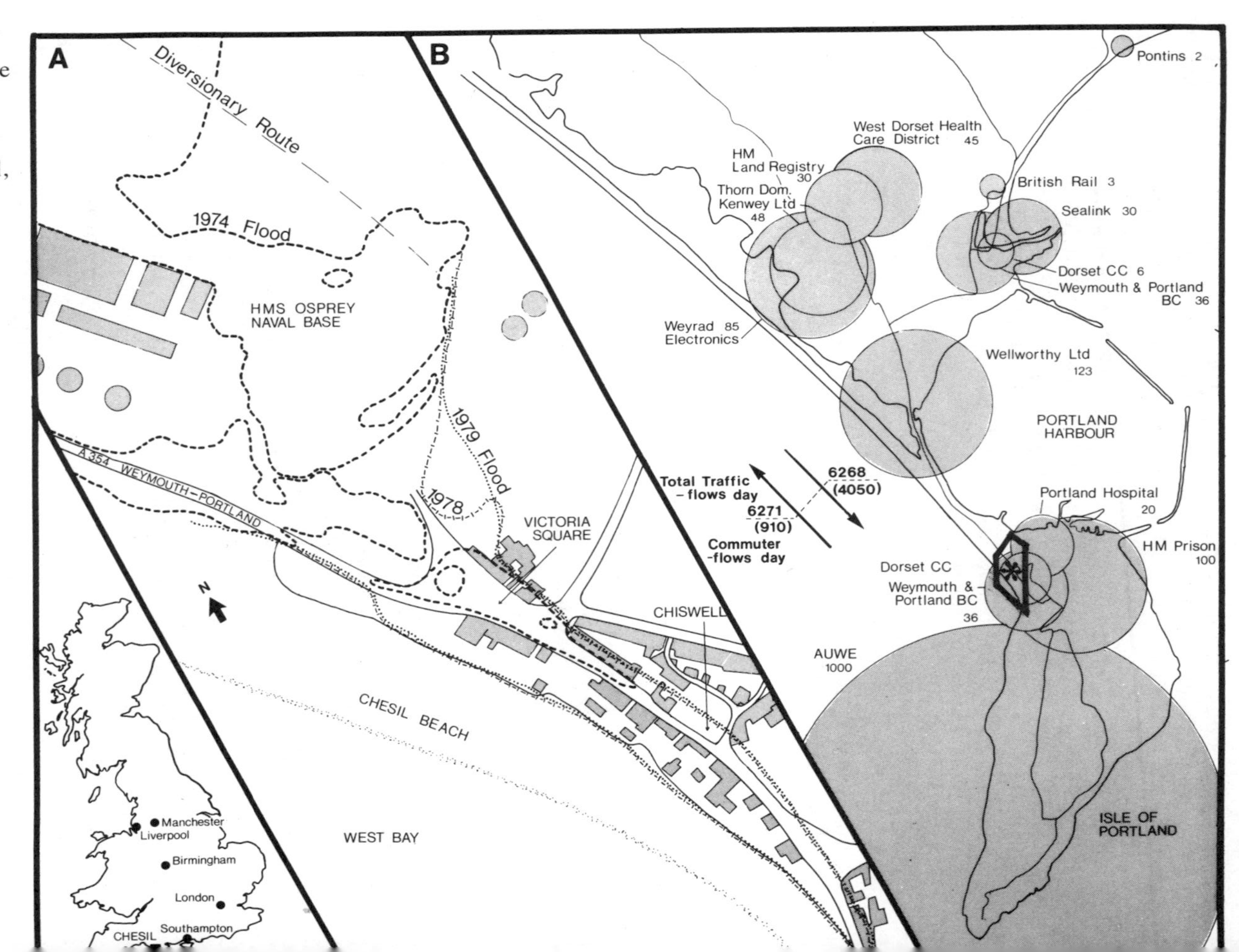

Figure 4.8 The flooded area and the extent of indirect flood effects on industry (shaded circles) for Chiswell, Dorset.

the evaluation of likely future flood damages, and thereby the benefits of flood alleviation, given the sparse recorded history of flood damage and urban encroachment of the flooded area in the recent past. Standard data were used to gauge basic flood damage with a 10–20 per cent supplement for the extra damaging effects of sea water, which research has attributed mainly to the extra household inventory damage from saltwater flooding (Penning-Rowsell 1978, Cole & Penning-Rowsell 1981). Site surveys of the structural damage to many buildings caused by the wave action were also necessary (Table 4.5); damage to cars was substantial in two major floods and this was assessed as their full second-hand value.

Table 4.5 Summary of direct and indirect flood damages at Chiswell, Dorset (April 1980 prices) (from Penning-Rowsell & Parker 1980).

		Percolation flood (£)	Storm surge flood (£)	Ocean swell flood (£)
(a)	Direct damage			
	extra damage to buildings	—	1200	79 600
	damage to public utilities	—	19 500	32 700
	damage to cars, etc.	—	4050	83 000
	Royal Naval facilities	—	10 000	20 000
	standard damage totals	6647	263 741	263 741
(b)	Indirect damage			
	cost of employment loss	72 716	145 432	290 864
	cost of mail delays	—	124	124
	cost of traffic disruption	18 515	37 031	74 062
	estimated loss of business profits, Chiswell	2360	9442	14 161
	reported costs of emergency services etc.	—	35 394	64 794
(c)	Totals	100 238	525 914	923 046

Significant indirect damage arises from the disruption during floods of the causeway linking the Isle of Portland to the mainland necessitating a radically new approach to evaluation (Penning-Rowsell & Parker 1980). The main loss is to local employers (Fig. 4.8) since some 3000 employees cannot get to work during flood events. In general these employees are paid on such occasions but no work is undertaken. Therefore the full amount of such a loss to employers in any future events was interpreted as constituting a benefit of flood alleviation, although recent analysis has modified this position to give a somewhat different measure of the value added by labour (see Green *et al.* 1983b). Similarly the lost profit on reduced trade in the village shops during flood conditions is an allowable benefit as are the marginal costs of emergency services provided by the Police, Wessex Water Authority, the Royal Navy and the Weymouth and Portland Borough Council (Table 4.5).

The full discounted benefits of protection up to the storm surge severity of flooding totalled an estimated £12.1 million, excluding the very considerable intangible effects. These comprised stress and anxiety for the local inhabitants, the loss of one life in 1978, and a marked deterioration of the village community which was partly attributable to the flooding problem. The benefit–cost ratio for the scheme at the 5-year return period approximated 2.1 : 1 and Wessex Water Authority began the necessary construction work in 1981. A higher design standard would still have shown a benefit–cost ratio greater than 1.0 but such a standard was technically problematic, would have been vastly more expensive for only marginal increase in benefits, and would have had a major impact on Chesil Beach as an SSSI.

The economic evaluation of agricultural land drainage

Protecting agricultural land from flooding and improving its natural drainage may require expenditure on field underdrainage, drainage ditches, embankments, pumping stations and improved arterial drainage (Ministry of Agriculture, Fisheries and Food 1974–7). To offset these costs two types of benefit should result from the works (Penning-Rowsell & Chatterton 1977, 1980). First, a greater economic return should be obtained from all agricultural uses of the area drained. This benefit is termed agricultural enhancement and it can result from increased crop and livestock yields – and hence perhaps returns – from an existing agricultural enterprise, or from a change of land use from that prior to the drainage, or from a mixture of both (Local Government Operational Research Unit 1978). Secondly, any crop damage caused by flooding should be reduced or eliminated (McDonald & Ledger 1981). In Britain the majority of agricultural drainage benefits fall in the first category since few farmers grow high-value crops in areas liable to serious flooding or suffering from significantly impeded soil drainage.

With the rapid increase in agricultural drainage since the Land Drainage Act 1930 (Green 1979), and the consequent diminution of Britain's wetland habitats, greater attention has focused on the economics of drainage and on the relative merits of drainage *vis-à-vis* conservation (Hinge & Hollis 1980, Penning-Rowsell 1983a, Royal Society for the Protection of Birds 1983). This attention has highlighted considerable problems in quantifying the economic benefits of drainage (Black & Bowers 1981, Bowers 1983). These problems require serious attention but this should not detract from the aim of using some form of economic evaluation to gauge worthwhile investment levels.

A FRAMEWORK FOR EVALUATION

The framework for evaluating the agricultural benefits of land drainage is not as well defined as that for urban flood alleviation (see Fig. 4.1). Nevertheless a number of authors and organisations, including the Ministry of Agriculture, Fisheries and Food, now appear to agree – but others do not – that certain parameters and procedures are of central importance. These include the use of gross margins, which measure the gross return from agricultural enterprises such as wheat or dairy cows before the deduction of variable costs such as fertilisers and casual labour. In general the higher the productivity of the land the higher will be the gross margin. Other procedures include allowing for the likely increase in fixed costs by way of farm investment needed to realise the benefits, the need to assess secondary benefits away from the immediate area of drainage, and the need to discount both costs and benefits to give their present values. Many aspects of the assessments, however, remain uncertain and controversial.

The Ministry of Agriculture, Fisheries and Food (1974, 1978) has indicated that the benefits of drainage and flood protection should comprise the difference between the total gross margin of the enterprises of the affected area without the installation of a drainage scheme and that accruing after a scheme is implemented and fully operational, less any increase in fixed costs. With this approach, where the farmer continues with the same livestock or arable enterprise after the scheme is implemented, the gross output from that enterprise will rise when yield per hectare rises as the enterprise benefits from the improved conditions. The variable costs such as seed, fertiliser or animal care costs should remain approximately constant so that the increase in gross margin should measure the increased enterprise return (Penning-Rowsell & Chatterton 1984).

Where there is a change in the type of agricultural enterprise, perhaps from low intensity grazing to intensive horticultural crops, the situation is more complex. Here what is exchanged is the low gross margin from the grazing for the higher margins from the intensive cropping. Naturally the farmer may well incur increased fixed costs to obtain this higher gross margin, perhaps from purchasing essential equipment such as a combine harvester or from employing more staff. Alternatively there might be loss of staff and capital released from the sale of equipment such as milking machinery if the change is from dairying to arable farming. Therefore the difference in gross margin with and without drainage needs to be corrected by the alteration in fixed costs so that a realistic and relevant benefit figure is obtained.

This view of the benefits of agricultural land drainage can be summarised as follows:

$$B = \sum_{t=0}^{t=n} \left\{ \frac{(ge_t - gp_t) - f_t - d_t}{(1+r)^t} \right\}$$

where B is the total benefit, t is a year during the scheme's life, n is the expected life of the scheme, ge_t is the expected crop or livestock gross margin in year t with flood protection and/or drainage, gp_t is the crop and livestock gross margin in year t without the scheme, d_t is the expected value of reduced crop damage in year t, f_t is the net change in farmer's fixed costs required to obtain the increased production in year t, and r is the discount rate (i.e. 0.05 = a 5% discount rate). This formula yields the capital sum to be weighed against scheme costs, without allowance for subsidies within either costs or benefits.

METHODS OF SURVEY AND ANALYSIS

Whatever theoretical framework is adopted to gauge the economic value of agricultural drainage – and the above analysis is certainly controversial – a comparison is necessary between the economic circumstances with and without the land drainage scheme. As such these assessments involve predictions – some would say guesses – about the future. Inevitably therefore they cannot be fully accurate and, indeed, should not pretend to be so. However, it is a matter of judgement how detailed is the analysis. A highly detailed assessment of one parameter – perhaps the market prices for livestock (Kavanagh & Slater 1975) – may be counterproductive when set against the possibility of only broad estimates of likely future stocking patterns and the uncertainty over gross margins in unstable agricultural markets.

Whatever method of survey is used, therefore, some generalised assessment is needed of future returns from agricultural enterprises. This assessment can be based on professional opinion of the most likely potential use for the drained land by Ministry of Agriculture, Fisheries and Food staff or specialist consultants. Alternatively, an assessment can be based on the judgements of the farmers involved, or using a combination of these methods.

Some form of farm survey is essential, however, to determine the area of the farm affected by flooding and waterlogging and also the area which will benefit. The latter may well not correspond exactly to the former. Areas away from the floodplain will benefit as farmers re-arrange their farm management system in response to the newly drained land. An example is the case of the proposal to drain the Amberley Wild Brooks in Sussex (Parker & Penning-Rowsell 1980, Hall 1978). Here more than 30 per cent of the benefits of this scheme arose from circumstances where farmers would be able to intensify grazing on floodplain land and so release upland areas for arable cultivation. These upland areas are traditionally needed to supplement the poor and uncertain grazing in the flood-prone Brooks but if these were drained and the grazing made more reliable the upland areas could be used more intensively and hence more profitably.

In addition to estimates of post-scheme gross margins, present and likely future economic returns from the undrained locations are required. Two approaches are possible here. Either a detailed financial analysis can be undertaken for the farms involved, or a more generalised assessment can be undertaken by determining the land uses of the areas affected and the stocking densities and types. To these data can be applied average gross margins to indicate approximate gross returns (Nix 1980). The latter approach is commensurate with the generalised assessments of future returns, although this may compound errors by using two sets of imprecise data (Penning-Rowsell & Chatterton 1980, 1984). Broad assessments of economic viability could perhaps precede a more detailed analysis if worthwhileness is in doubt.

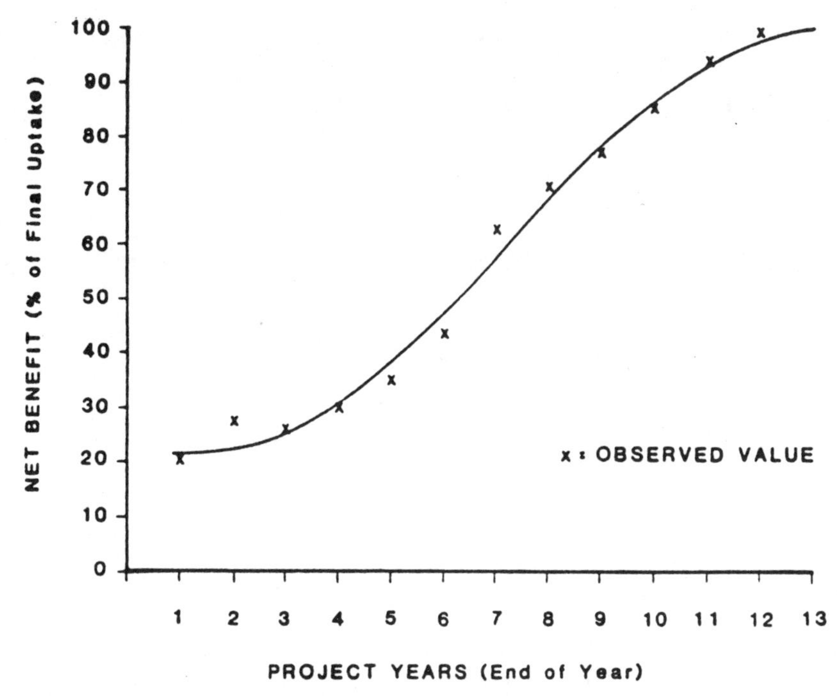

Figure 4.9 Predicted agricultural benefit uptake curve (from Morris & Hess 1984).

To arrive at a capital sum worth spending on the drainage works involves two further operations. First, the benefits of drainage in the form of increased gross margins will not occur immediately. The farmer is unlikely to make the necessary changes to his farming practices until the scheme has been implemented and he or she is confident of its success in regulating the water table. Thus there will be a period of 'uptake' of benefits as farmers install the necessary underdrainage in their fields and make other changes to their farming systems. Recent research (Fig. 4.9) shows that after an initial surge such uptake will generally be slow after the scheme is completed, then accelerate, and then slow again as all the worthwhile underdrainage is completed (Morris & Hess 1984). The process may take decades in the case of a large scheme involving many small farms and perhaps 5 years even if only a small number of farmers is involved, each of whom is favourable to change, and has institutional support such as from an Internal Drainage Board.

Secondly, the benefits have to be discounted to give their present values (p. 100). The result is the maximum capital sum worth spending, or investing, to obtain the future benefits, and this sum can be compared with the scheme costs incurred by the farmers, Internal Drainage Boards, County Councils, Water Authorities and the Ministry of Agriculture, Fisheries and Food.

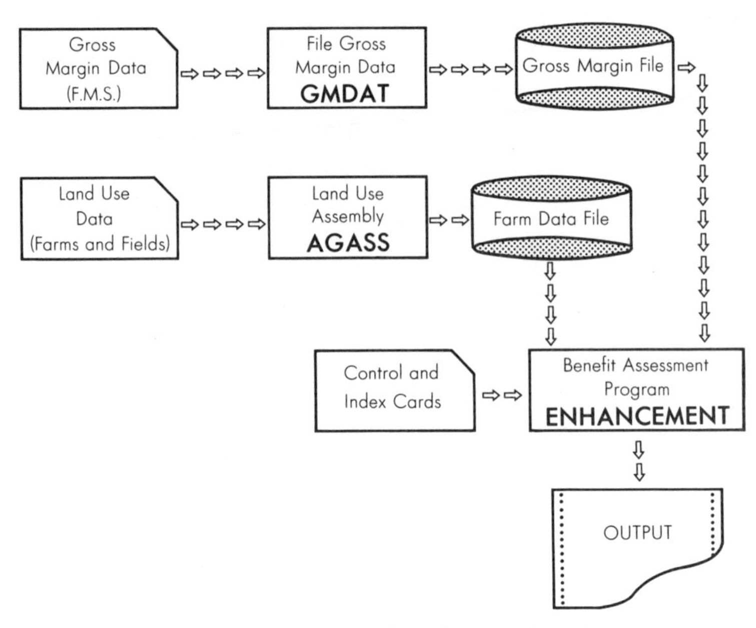

Figure 4.10 Computer 'model' for assessing the benefits of agricultural drainage (from Penning-Rowsell & Chatterton 1977).

AN ASSESSMENT MODEL

A computer model has been developed to systematise assessing the benefits of agricultural drainage (Fig. 4.10). The model allows comprehensive testing of the sensitivity of the results to the many significant assumptions involved.

Gross margin data appropriate to the benefit assessment in question are assembled by a program called GMDAT which selects data for the types and sizes of farms in the scheme area. The appropriate gross margin data are linked with the livestock enterprise or land use of the field or block of land, which is assembled by program AGASS. Current gross margins are thereby calculated.

Likely post-scheme gross margins are calculated from farmers' estimates of future farming systems following drainage. Agricultural enhancement is calculated as the difference between these two. Within this process gross margins can be adjusted in a number of ways, including for the locations of the land (e.g. whether above or below the Medway Letter Line) and for the gross margin's subsidy or support element.

Finally the model calculates the phased uptake of benefits, given data on likely local uptake rates, and discounts benefits to present values. A full discounted cash-flow analysis is the most elegant way of presenting the results (Fig. 4.11). This shows the stream of annual benefits compared to capital and recurring maintenance costs, each discounted to present values, and allows the deduction of the nominal terminal capital value of the investment at the end of the scheme's design life (Chatterton & Lau 1983).

DISCOUNTED CASH FLOW ANALYSIS

	COSTS							BENEFITS								
								PHASED ANNUAL AGRICULTURAL			ANNUAL URBAN					
YEARS AFTER SCHEME	ARTERIAL	MAINTENANCE	UNDER DRAINAGE	FARM COSTS	O&M COSTS		PV OF COSTS	STAGE I	STAGE II	STAGE III	STAGE I	STAGE II	STAGE III	PV OF BENEFITS	CASH FLOW	PV FACTOR 5%
0	226765	0	0	0	0	0	226765	0	0	0	0	0	0	0	-226765	1.0000
1	282150	0	13340	0	0	0	281419	389	0	0	0	8997	0	8939	-272480	0.9524
2	327085	0	18595	0	0	0	313541	2481	1398	0	0	8997	6812	17857	-295684	0.9070
3	0	0	36382	0	0	0	31428	3381	8891	3325	0	8997	6812	27129	-4299	0.8638
4	0	0	0	0	0	0	0	3829	12112	21134	0	8997	6812	43507	43507	0.8227
5	0	3586	0	0	1214	0	3760	4069	13717	28792	0	8997	6812	48881	45121	0.7835
6	0	5596	0	0	1238	0	5099	4069	14577	32603	0	8997	6812	50039	44940	0.7462
7	0	4022	0	0	1262	0	3755	4069	14577	34648	0	8997	6812	49110	45355	0.7107
8	0	0	0	0	1288	0	871	4069	14577	34648	0	8997	6812	46771	45900	0.6768
9	0	0	0	0	1314	0	847	4069	14577	34648	0	8997	6812	44544	43697	0.6446
10	0	3586	0	0	1339	0	3023	4069	14577	34648	0	8997	6812	42423	39400	0.6139
11	0	5596	0	0	1367	0	4071	4069	14577	34648	0	8997	6812	40403	36332	0.5847
12	0	4022	0	0	1394	0	3015	4069	14577	34648	0	8997	6812	38479	35464	0.5568
13	0	0	0	0	1422	0	754	4069	14577	34648	0	8997	6812	36646	35892	0.5303
39	0	0	0	0	2380	0	354	4069	14577	34648	0	8997	6812	10306	9952	0.1491
40	0	3586	0	0	2428	0	854	4069	14577	34648	0	8997	6812	9815	8961	0.1420
41	0	5596	0	0	2477	0	1092	4069	14577	34648	0	8997	6812	9348	8256	0.1353
42	0	4022	0	0	2526	0	843	4069	14577	34648	0	8997	6812	8903	8060	0.1288
43	0	0	0	0	2577	0	316	4069	14577	34648	0	8997	6812	8479	8163	0.1227
44	0	0	0	0	2629	0	307	4069	14577	34648	0	8997	6812	8075	7768	0.1169
45	0	3586	0	0	2680	0	697	4069	14577	34648	0	8997	6812	7690	6993	0.1113
46	0	5596	0	0	2734	0	882	4069	14577	34648	0	8997	6812	7324	6442	0.1060
47	0	4022	0	0	2789	0	687	4069	14577	34648	0	8997	6812	6975	6288	0.1009
48	0	0	0	0	2845	0	273	4069	14577	34648	0	8997	6812	6643	6370	0.0961
49	0	0	0	0	2901	0	265	4069	14577	34648	0	8997	6812	6327	6062	0.0916
50	0	3586	0	0	2960	0	570	4069	14577	34648	0	8997	6812	6026	5456	0.0872
TOTALS	836000	122422	68317	0	90226	0	918366							1107122		
TERMINAL VALUE	0	0	0	0	0	0	0									
TOTALS	836000	122422	68317	0	90226	0	918366									

BENEFIT/COST RATIO = 1.206 NET PRESENT WORTH = 188756

Figure 4.11 An example of a discounted cash-flow analysis of both costs and benefits of flood alleviation (from Penning-Rowsell & Chatterton 1984).

THE PROBLEMS

The Ministry of Agriculture's view of benefit assessment, as discussed above, has changed over time as new insights into land drainage economics have developed. However, certain problems undoubtedly remain (Penning-Rowsell & Chatterton 1984). For example, insufficient data are available on take-up rates, on the gross margins obtainable in flood-prone areas and other poorly drained land, and on the future prices for agricultural products, all of which directly affect the calculated benefits of drainage works. Furthermore, the farm interview survey data could suffer from the 'free rider' effect (Bowers & Black 1983). This hypothesis suggests that respondents will exaggerate their responses when others would pay the consequences, in this case the general taxpayer contributing to a drainage scheme, although no systematic free-riding has been demonstrated (Stroebe & Frey 1982).

However, undoubtedly the major problem concerns the prices of agricultural products. The central point is that when public funds are used to obtain agricultural improvement it is the social benefits of these schemes that should be compared with the costs, rather than the increased private financial returns to the individual farmers (Local Government Operational Research Unit 1978). Part of the price obtained by the farmer at market for his or her produce comprises subsidies (Black & Bowers 1981). Prices are also artificially raised by agricultural support policies, which involve intervention buying at guaranteed minimum prices (Body 1982). The gross margin used to measure the benefits of agricultural drainage includes these subsidy elements, which are in fact payments to farmers by the state and are not real value-added resources (Bowers 1983).

Because they contain subsidies, unadjusted gross margins should not be used to gauge the benefits of flood alleviation and agricultural drainage. One alternative is to use world prices for the commodities in question, although these can be too low owing to 'dumping' at below true cost. Another is to adjust gross margins to remove their subsidy component, although this is difficult. Nevertheless, the implications of using adjustment factors to account for protection and subsidies is that some values for gross margins may be very low or even negative. In this case much agricultural drainage undertaken at present may not be worthwhile when real resource creation is the criterion. At its most crude, it may not be worthwhile for the nation to contribute to draining land only for crops to be grown which are subsequently bought at intervention prices and disposed of at a loss! There may well be political reasons for maintaining drainage programmes, despite their poor economic performance, in the interests of greater food production. However, this decision should be clear rather than obscured by apparently satisfactory economic analyses which are at least partly weighing private gains against social costs.

INTANGIBLE EFFECTS

The intangible effects of agricultural drainage can be considerable. In the main these comprise the adverse effects of drainage upon wildlife and landscape values (Ch. 5). By definition these effects cannot be given monetary values for inclusion in the cost–benefit analysis; nevertheless they should be seriously considered. The most appropriate way that intangible aspects can be included in decision making is to compile lists of intangible effects of particular floods, flood alleviation schemes and land drainage proposals (as in Table 4.3). In this way the economic gains or losses from the schemes being considered can be weighed against, or with, the important but as yet unmeasurable effects.

The costs of flood alleviation and land drainage

The costs of flood alleviation or agricultural drainage will include the value of the materials used, such as concrete, steel or electronic equipment for a warning scheme, together with the labour costs of design, construction and maintenance (Institution of Civil Engineers 1969). The organisation concerned with flood alleviation may have to borrow the money to finance operations and thereby incur costs as interest charges. These charges do not, however, affect the cost–benefit calculations. They are simply transfers between members of the community and do not represent real use of resources; they are a financial liability for the borrower which is exactly matched by the financial gain by the lender.

Maintenance and other recurring costs, perhaps for replacing weirs or sluices during the life of a scheme, have to be discounted to present values to allow appropriate comparison with scheme benefits. Discounting in this context should encourage high maintenance/low capital cost schemes, perhaps of the non-structural variety, since future expenditure on operation and maintenance will not feature largely in discounted costs. However, most land-drainage organisations are concerned not to incur heavy operation and maintenance liabilities owing to their effect on revenue and hence their charges.

The investment in drainage works should be valued net of taxation, such as indirect and fuel taxes, and also at its opportunity cost. This is the value within the economy of the alternative opportunities for the material, labour and services used. Most controversial here is the valuation of labour, which forms a high proportion of drainage expenditure. In theory, the opportunity costs of labour at times of long-term structural unemployment can be considered to be close to zero in that it is not productively employed in the economy before working on drainage schemes (Green & Penning-Rowsell 1983). Alternatively, one might value this opportunity cost as the difference between wages and unemployment benefit.

Such arguments advocating shadow wage rates are not accepted by the British Treasury (H.M. Treasury 1982) since they consider it inadmissible to seek macro-economic corrections to the economy as a whole via micro-economic valuations of project investment. If this policy ever changed it could profoundly affect the calculated worthwhileness of flood alleviation and land drainage (but also of motorways, hospitals and all other public investment).

The Soar Valley Improvement Scheme

The Soar Valley between Leicester and Nottingham has a history of flooding, mainly affecting agricultural land. The valley's drainage has remained unimproved because of a complex division of legal responsibilities. This occurs largely because the river Soar and the associated canal merge and separate several times down the valley. However, the Severn Trent Water Authority and its predecessors have no powers under the Land Drainage Acts to alter the navigation channel and thus a complex Parliamentary Bill was required to obtain the necessary powers.

This Bill was almost passed unopposed in 1982 until the Council for the Protection of Rural England (CPRE) persuaded the House of Lords to establish a special Select Committee to investigate both the environmental effects of the proposed scheme and its economic justification. This opposition should be seen as part of the long-standing and penetrating critique by the CPRE of the cost–benefit analysis of land drainage schemes (Hall 1978, Bowers 1983). The critique is part of a campaign designed to

Table 4.6 Soar Valley Improvement Scheme: comparison of Ministry of Agriculture approved appraisal results with an economic evaluation with partial shadow-pricing of labour, other costs adjusted for their taxation element, and benefits valued at world prices (Chatterton 1983).

	Present value (£)*		Net present worth	Benefit–cost ratio
	costs	benefits		
quasi-financial appraisal†	7.282	9.400	2.118	1.29
economic appraisal	5.873	8.237	2.364	1.40‡

*Discount rate 5 per cent; scheme construction period 7 years; costs include design and supervision costs which would probably be incurred by Severn Trent Water Authority irrespective of constructing the Soar scheme.

†In accordance with Ministry of Agriculture, Fisheries and Food (1974, 1978) guidelines (costs at tender prices; benefits at farm gate prices, etc.).

‡The principal reason why this ratio is higher than 1.29 is that farmers are seeking to convert to arable cultivation, for which world prices were relatively higher than for their pre-existing dairying. Costs net of fuel tax and other transfer payments thus reduce more than do the benefits.

reduce or halt the landscape changes that these schemes produce.

The Soar Valley scheme proposed improved arterial drainage from which some 2795 hectares would produce higher gross margins contributing to a total agricultural benefit of £6.58 million, together with £0.94 million urban benefits and £1.88 million from the alleviation of traffic disruption. The Severn Trent Water Authority, contrary to the Ministry of Agriculture, Fisheries and Food's (1974, 1978) recommendations, revalued the benefits by reducing the agricultural valuations to world prices, thus eliminating the effects of UK and EEC subsidies. The costs, however, were also reduced to eliminate taxation and other transfer payments and also by some shadow pricing of labour inputs (Severn Trent Water Authority 1983a). The results (Table 4.6) show that the scheme is worthwhile on both assessments, as judged by comparing tangible costs with tangible benefits. The House of Lords Select Committee (1983) supported the use of the 5 per cent discount rate and, implicitly, the Ministry of Agriculture, Fisheries and Food's (1978) controversial benefit assessment guidelines. In their report the Committee ignored the critical questions of agricultural prices, and the shadow pricing of some costs, and approved the Bill in July 1983. No doubt the CPRE will return to the inadequately aired issues on some future occasion.

Assessment

Economic evaluation of flood alleviation and agricultural drainage schemes presents many conceptual and technical problems. These cannot be ignored and should form the basis of further research.

Economic evaluation is merely the tool of the planner of flood alleviation and drainage schemes. Political considerations resulting in public pressure or sectional interests will always be important to decision making which can therefore never become the mechanistic 'linear deductive' process implied by Figure 1.7. However, the aim of economic evaluation of ensuring optimal use of scarce resources should not be lost. There is this danger if engineers use economic techniques simply to rationalise decisions already taken, perhaps to protect against flooding or waterlogging to an arbitrary design standard. The opposite approach is a better way. Here the analysis of benefits is done first and then schemes are designed to invest as little as possible to obtain these benefits.

With this approach alternative ways of alleviating the flooding should enter into consideration more readily. These alternatives might include whether a warning scheme is more appropriate than structural flood control to save the expected flood damage. Some other form of agricultural improvement such as greater use of fertilisers or improved labour productivity might secure more economical increased food production than investment in land drainage. At the very least economic evaluation may prompt examination of alternative standards for conventional schemes to see if a more or a less ambitious design provides better value for money. Used in this way economic evaluation can pose fundamental questions concerning the use of public investment for greater community welfare rather than remaining just a weapon in the hands of those seeking to restrain public expenditure.

5 *Environmental impacts and conservation policies*

The growing significance of environmental impacts

Both urban flood alleviation and agricultural land drainage schemes can have significant adverse and unintended environmental effects (Fig. 1.2). Agricultural land drainage, in particular, can radically impoverish many rich and varied wetland habitats. Rare plants and animals which require wetland conditions for their survival can be destroyed, and drainage can irreversibly change landscapes to which there is an enduring attachment.

Within the long history of land drainage in Britain the awareness of and concern for these environmental impacts is quite recent. Hinge and Hollis (1980), Hollis (1980) and Newbold (1982) show that few river engineers before the mid-1970s were concerned for the environmental impact of their work. Few concerns were expressed at this time over the impact of land drainage upon nature conservation values (Nicholson 1972). More emphasis was put then in the vanguard of environmental opinion and research on the deleterious effects of pesticides and the removal of key agricultural habitats such as hedgerows.

It is only with the increasing rarity of wetland sites following decades of active drainage that the value of those remaining has risen, both in Britain (Green 1979) and overseas (Green 1980, Hill 1976). The issue has become of major concern to government organisations such as the Nature Conservancy Council (Newbold *et al.* 1983) and voluntary bodies including the Royal Society for the Protection of Birds (1983), The Council for the Protection of Rural England (1983) and The Society for the Promotion of Nature Conservation. Only recently, however, has there emerged a response to this concern, with professional engineering recommendation moving away from flood-alleviation channels designed purely for hydraulic efficiency and towards incorporating aesthetic and wildlife considerations (Water Space Amenity Commission 1978, 1980a,b, 1983). This change has been prompted by pressure from amenity organisations and a lead from a small band of enlightened engineers (Miers 1967).

In this chapter we assess some of the effects of drainage upon flora, fauna and aesthetic values. However, the environmental impact of land drainage is highly complex although certain overriding considerations stand out. First, the specific impacts of individual schemes for flood alleviation or agricultural drainage depend crucially upon local circumstances. Secondly, there is therefore the need for proper research and for comprehensive surveys of the nature and extent of wetland habitats and thereby the potential environmental impact of specific drainage schemes. Only then can a meaningful assessment be made of the intangible adverse aspects of flood alleviation and drainage with which to complement the cost–benefit analysis of measurable economic effects (Ch. 4). Thirdly, certain common principles are relevant to the design and management of many drainage works so as to minimise conservation loss and, fourthly, there is the need for extensive consultations between the interested parties before drainage authorities proceed to make what may well be irreversible decisions (Ch. 6).

Nature conservation and amenity values

THE COMPLEXITY AND TOTALITY OF IMPACTS

Living organisms and their non-living (abiotic) environment are inseparably interrelated and interact upon each other. This is a central concept or principle in ecology (Odum 1971): one change in any element of an ecosystem is liable to affect the total system. Every species has adapted and evolved to fill a particular ecological niche and has a complex series of interrelationships with surrounding species and physiographic features.

If a tree is therefore cut down to allow easier channel clearance, the insects and birds that live on it will also be affected, which in turn may affect other animals that depend on those insects for food or were eaten by birds. The environmental impact of drainage work may therefore spread from the immediate site to neighbouring interconnected ecosystems, either wetland or otherwise. This spread is analogous to the way drainage for improved agriculture can influence the management and economics of the whole farm affected or, alternatively, flood protection for a factory may ensure employment and thereby influence the prosperity of the wider local economy. The important implication of the ecological interrelationships, however, is that change may be self-perpetuating to produce radical effects from quite small initial disturbances to water regimes.

When a wetland is drained or a river is realigned there are direct effects in the form of loss of habitats. Open water usually disappears, thus eliminating the habitat of many bird species (Royal Society for the Protection of Birds 1983). The management of drainage ditches and remaining other wet areas within an improved agricultural regime is too intense to support the aquatic plants and other wildlife usually found in semi-natural wetland conditions (Newbold 1982). Modification of ditches, streams and river channels to promote drainage in both urban and rural areas alters channel form, discharge, water temperature and chemistry (Fig. 5.1).

Drainage also brings indirect effects. Agricultural intensification usually means greater use of fertilisers, herbicides and pesticides which run off into ditches and channels and so reduce the density and diversity of aquatic organisms (Hill 1976). Channel improvements to pass flood peaks through urban areas usually require regular maintenance, which may inhibit the recolonisation of wildlife.

Drainage works can also have secondary effects beyond the immediate drainage site. More than any other habitats wetlands are sensitive to environmental changes outside their boundaries (Gilman n.d.), particularly the drainage of surrounding areas (Hill 1976). The edges of undrained areas adjacent to drained land will show a successional change as they dry out under the influence of lowered surrounding water table levels. Moisture-

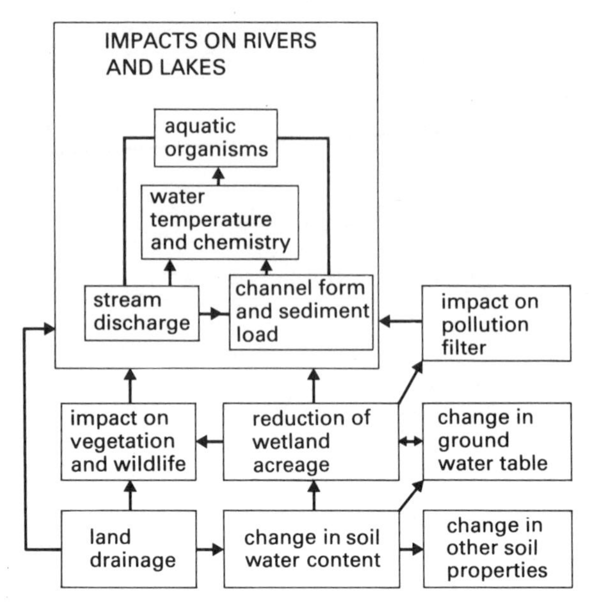

Figure 5.1 Environmental impacts of agricultural land drainage (from Hill 1976).

loving plants will give way to species adapted to drier conditions and eventually colonisation by scrub or trees could replace the marshland species. The implication of all these effects and interdependencies is that a particular species or habitat cannot be protected in isolation without also conserving the complete environmental system in which it exists.

CONSERVATION VALUES

Conservation can be defined as seeking 'to manage human use of natural resources for the greatest benefit of present and future generations. To many it has not only a scientific, but also a philosophical or ethical base, with strong emotive appeal' (Water Space Amenity Commission 1980b, p. 13). Most wetland areas are far from natural but depend for their character on a complex and traditional type of management for a modest economic return. The rationales for their conservation in this state are many and controversial. The protection of endangered habitats will retain their scientific and educational value; there are moral and ethical reasons for human beings not destroying other species. Existing 'unimproved' landscapes may give aesthetic and amenity value for recreation and even commercial return for sport and tourism. Many arguments in this field revolve around maintaining for future generations the environmental diversity and opportunities we currently enjoy. This requires a longer term view of environmental use than is commonly current in our society today.

Wetland conservation is complex, partly because there are many different types of wetland habitat and landscape (Table 5.1). Not all are equally valuable and not all are equally sensitive to land drainage or flood alleviation schemes. The ecological processes operating are themselves very

Table 5.1 A simple classification of wetland types (from Water Space Amenity Commission 1980a).

Upland or lowland	Lowland	Coastal
lakes	marshes	estuaries
ponds	fen	salt marshes
rivers		
streams		
bogs		

complex, given that these habitats are dynamic open systems continually affected by external forces such as changing water characteristics or grazing intensity. The natural process of ecological succession will continually occur and to maintain the *status quo* requires active management, as particularly exemplified at Woodwaltham Fen (Duffey 1974). Here the whole SSSI depends on the fen being separated by clay banks from the surrounding drained area and thus maintained with what today is a completely artificial hydrological regime. In general natural wetland succession, ultimately from open water to woodland carr, results in a declining diversity of species rather than an enrichment. Much of the management at Woodwaltham has thus aimed at controlling the spread of carr and protecting the now almost unique assemblage of mixed sedge fen (Sheail & Wells 1983).

Decisions not to drain specific wetlands but to manage them for their nature conservation or landscape significance mean implicit or explicit valuation of their characteristics. Economic evaluation of food with market prices is relatively straightforward, but conservation value is much more problematic. Such value can be a function of many factors including diversity, naturalness, rarity, the geographical size of a resource, the typicality or representativeness of a site, and its fragility or geographical position (Ratcliffe 1977, Newbold *et al.* 1983). Some of these factors are incompatible: the protection of rare species may necessitate limiting diversity to maintain a particular ecological niche. A fragile site is unlikely to be useful as a type example, and so on.

The problem is compounded because different specialists emphasise different factors or facets of value. The botanist might value highly a species-rich meadow which to an ornithologist is unimportant as it supports only common birds. In ranking sites some botanists would stress species rarity rather than a diversity of the less uncommon plants. For many people naturalness contributes uniquely to value, whereas others can rank equally the natural and created environments. There is thus neither a clear professional scientific nor a popular consensus on what is indisputably valuable. In the same way there are different landscape tastes and there is no simple consensus on what is attractive or otherwise scenically important. Nevertheless, despite this somewhat blurred vision of what should be protected and what could be sacrificed, decisions are continually being made on the design of both urban flood alleviation and agricultural land drainage schemes which have significant and excessive adverse environmental effects.

Urban flood alleviation schemes

Rivers and streams are often among the few places where the urban population can experience some of the tranquillity of apparently natural areas and where wildlife can maintain a foothold in what is otherwise an unfavourable environment. There is thus both aesthetic and ecological potential in such locations; the public may appear indifferent to such considerations, but the engineer can lead in the promotion of better designs which the public subsequently may well come to value highly. It is certainly possible to design schemes that operate efficiently in engineering respects and which are also aesthetically pleasing and provide habitats for wildlife. The increased cost of such balanced design can be marginal (Newbold *et al.* 1983).

IMPACT ON HABITATS

In the past many urban flood alleviation schemes have been designed without consideration for habitat protection or enhancement. The result was a predominance of sterile, concrete-lined, box-shaped channels replacing habitats where at least some wildlife may have existed. Fish life in such concrete channels cannot be sustained at low summer flows, even if pollution levels permit, given the lack of water depth and aquatic plants. There is insufficient soil for plants to germinate and grow, and in any case these would be cleared periodically to maintain the channel's design flow capacity. Ward (1978) presents some revealing photographs of channels for the Pymms Brook in Edmonton, London, as part of the major river Lea flood alleviation scheme. The hydraulic efficiency is maximised at the cost of both visual interest and conservation value.

Many habitat improvements can result, however, from very simple modifications to the form of river and stream channels at minimal extra cost (Water Space Amenity Commission 1980a). Double-section channels provide both the concentration of low flows and the capacity to pass flood peaks, but fish are still left in summer months with pools at least 0.6m deep. Thus designed the river channel is self-cleansing in that concentrated low flows still remove litter and provide oxygenation of the water. The channel can be varied in plan form thus creating habitat diversity including perhaps semi-protected islands for plants and wildlife sanctuaries. As water quality improves, with further investment in water reclamation, more plants or wildlife can be reintroduced and further semi-natural habitats created within urban areas so that fish, birds and other wildlife may flourish.

The routine maintenance of flood relief and other urban drainage channels similarly requires an understanding of habitat conservation. Channels should not be cleared wholesale or else plant species cannot regenerate and wildlife will be lost. Regular minor maintenance is preferable to infrequent major clearance, and major equipment needs careful use to avoid damaging the features such as pools and the shelter of riverbank vegetation that provide wildlife with the important ecological niches essential to their survival.

IMPACTS ON AMENITY AND RECREATION

Aesthetically pleasing shapes can be incorporated into good design to provide both efficient drainage and visual harmony. The use of gentle curves for both the plan and the cross section of artificially modified channels appear to be more visually pleasing than sharp corners and rigid shapes. Using vernacular materials with local rocks, bricks and stonework relieve the boredom of uninterrupted concrete channels. Rock-filled gabion cages can promote the growth of plants and shrubs which soften the profile and plan of channel works. When trees and shrubs cannot be planted in banks, owing to their encouraging erosion, Miers (1979) shows how spoil heaps, fringe drains and berms have been successfully planted on land bought specifically for this purpose.

Advantage may be taken of channel realignment to provide extra parkland sites, walkways and popular recreation areas to capitalise on the invaluable amenity spaces that a river can provide in urban areas (Fig. 5.2). This requires a comprehensive landscape plan such as that for the river Brent in North London. This plan was designed both to enhance human recreation provision and to encourage the reintroduction of fish, animal and plant life into what was previously a degraded, rubbish-filled stream of little recreational or amenity value.

Well designed schemes need to be linked to other developments and thus take the opportunity to provide wider environmental improvement than those just affecting the channel works. For example, channel realignment may release land for housing or business development. The replacement of bridges may lead to traffic flow improvements. Balancing reservoirs to take flood peaks may be empty when not in use or they can be left partially filled and adapted for fishing, with shallow water at the edges to encourage aquatic vegetation. This has been achieved most successfully in the new town of Milton Keynes where a real recreation and amenity resource has thus been created within the design of the urban drainage system (Anglian Water Authority n.d., Kelcey 1982).

Agricultural flood protection and field drainage

Drainage in rural areas is designed to lower and control the water table to allow more intensive grazing or arable cultivation. The most serious effects on both wildlife habitats and landscapes occur where large schemes involve the initial drainage of non-agricultural land to bring this into cultivation as improved pasture or for arable cropping. For the period 1971–80 some 9.6 per cent of all the area drained in England and Wales comprised this 'new' drainage of unimproved wet grassland. In these conditions the objectives of and conditions for efficient drainage are diametrically opposed to those required for conservation of wetlands and their highly specialised ecosystems. Drainage in other areas, linked to modern agricultural practices, also tends to create blocks of monoculture thus limiting the maintenance and diversity of wildlife habitats (Nature Conservancy Council 1977) and the loss of significant landscape features (Blenkharn 1979, 1983).

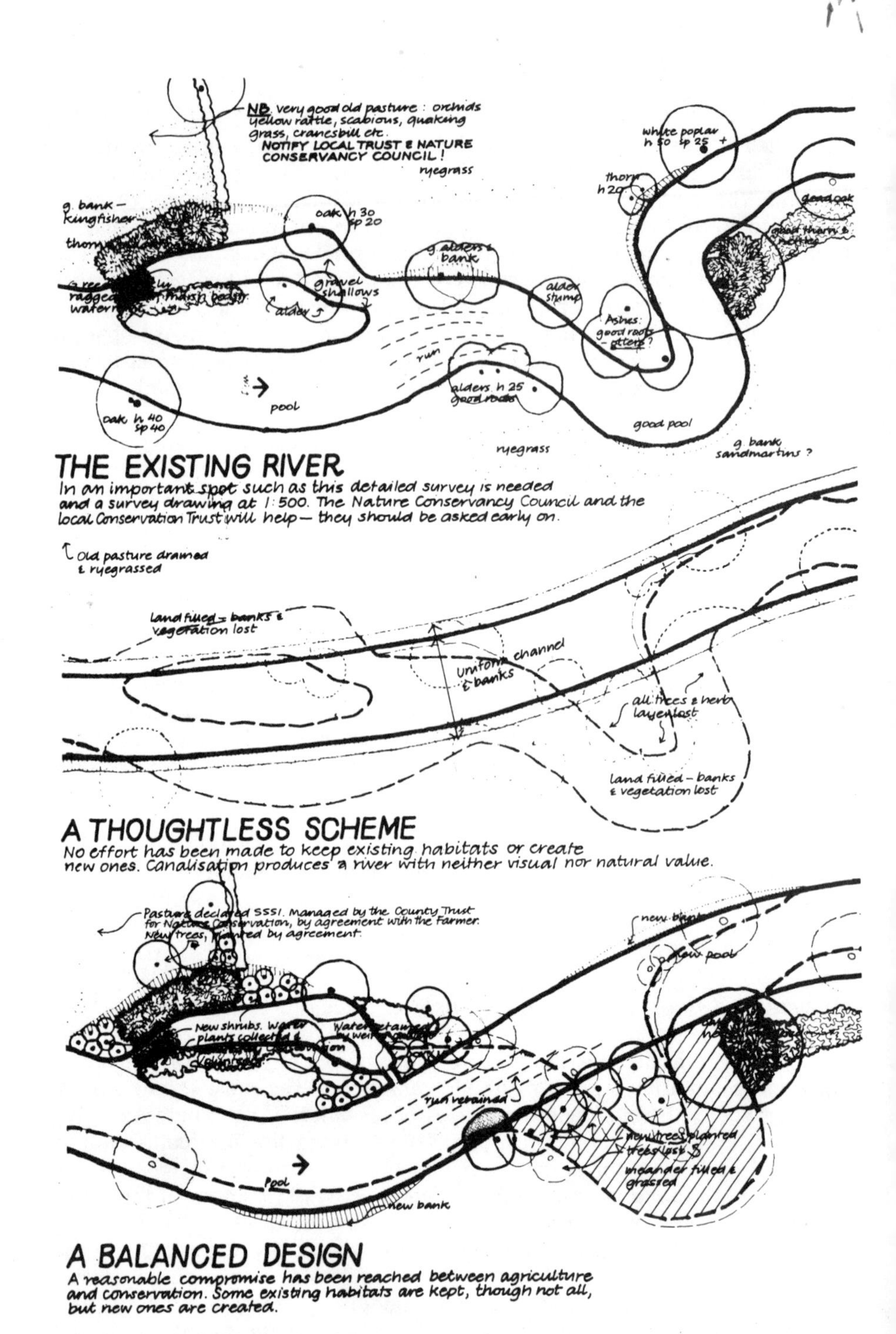

Figure 5.2 An example of amenity gain from sympathetic flood alleviation design (from Water Space Amenity Commission 1980a).

HABITAT LOSS: THE IMPACT ON WETLAND FLORA

The impact of drainage on the type of floristic richness characteristic of wetland areas naturally varies with habitat type (Table 5.2). When open water such as ponds are drained they will lose the floating plant species such as duckweed, water lilies, hornwort, the broad-leaved pondweed and the submerged rooted plants such as the water millefoil. The impact of channel works on rivers will remove bulrushes and bur-reed which are common in slow-moving streams.

Table 5.2 Types of wetland flora in Britain (from Haslam 1973).

Type	Common dominants	
	Latin name	Common name
(1) In or near ponds and lakes		
1.1 margins of open water	*Phragmites communis*	common reed
	Schoenoplectus lacustris	bulrush
1.2 silted ponds	*Phragmites communis*	common reed
	Typha latifolia	great reedmace
(2) In or near watercourses		
2.1 shallow lowland brooks	*Rorippa nasturtium-aquaticum*	watercress
	Sparganium erectum	bur-reed
2.2 shallow dykes, usually on alluvial plains	*Phragmites communis*	common reed
	Typha spp.	reedmace
2.3 fringes of rivers, canals, drains, etc.	*Glyceria maxima*	sweet reed grass
	Sparganium erectum	bur-reed
2.4 flood meadows beside watercourses 2.5 washlands	Vegetation varies with water table and use, from grass sward to *Phragmites communis*	
2.6 valley bogs of high nutrient status	*Carex paniculata*	panicled sedge
	Phragmites communis	common reed
(3) In large low-lying alluvial plains (in addition to 2.2, 2.3 and 2.4 above)		
3.1 on peat fen; fens, reedswamps	*Cladium mariscus*	saw sedge
	Phragmites communis	common reed
3.2 on mineral soil; marshes, reedswamps, etc.	*Glyceria maxima*	sweet reed grass
	Phragmites communis	common reed
(4) In coastal areas		
4.1 salty	*Spartina townsendii*	spartina grass
4.2 only slightly salty	*Phragmites communis*	common reed

Marsh drainage will destroy the stands of rushes and willow herb, dropwort and sedges growing where there is a soil with high organic content. Even on rough-grazed marshes these plants would survive within a mixed community of sedges, rushes and the coarse tussocky hairgrass, but drainage and intensive cultivation will eliminate the moisture-demanding species. Lowland marshes are usually characterised by trees such as the common alder, ash, willow and, in drier conditions on more acid peaty soils, the hairy birch and myrtle. Full-scale agricultural improvement will remove these and the cover or shade they provide for wildlife and specialised plant species.

The successional changes promoted by drainage affect the soil–water regime and thereby the conditions for plant growth. Underdrainage on impermeable alluvial soils means that winter rains permeate more rapidly. Prolonged waterlogging becomes rare and open water infrequent and shortlived. The soil above the drains becomes aerated and can absorb moderate rainfalls without becoming saturated. On permeable soils the effects of underdrainage depend on the cause of waterlogging. If this is due to rainfall the drainage will lower the water table and rain will not saturate the soil but will cause rapid variation in soil–water conditions. If waterlogging is controlled by ground water incursion the drainage will lower the water table and allow the rainfall to be absorbed rather than creating immediate runoff, so altering the chemistry of the water environment for plant species from ground water to rain water domination (British Trust for Conservation Volunteers 1976).

Drainage of alluvial meadows will encourage ploughing for arable cropping or intensified grazing. Both will reduce the floristic richness which is common on ground unploughed for generations. Examples of floristically valuable wet grasslands or permanent rough grazing meadows include North Meadow at Cricklade in Wiltshire and Port Meadow in Oxford. These areas provide habitats for such rare species as adders' tongue and the snakeshead fritillary which will not survive drainage and the consequently increased grazing density. Prior to 1930 the snakeshead fritillary was to be found in 116 10 km squares in 27 English counties; in 1970 it was only present in 15 of such squares (Royal Society for the Protection of Birds 1983).

Acid bog drainage will eliminate sphagnum mosses, heathers and the many specialised small plants including sundews and the many lower plants such as mosses and liverworts. The smooth mossy bog surface is replaced by an irregularly tussocky surface with sheathed cottongrass and heather. Bog mosses become increasingly rare and, with drier conditions, certain new mosses and liverworts appear which can take advantage of decreased competition or the oxidising peat but the resultant range of species will be much depleted.

Fen vegetation, as with most wetland ecosystems, is not natural and different types of fen result from different management regimes. Each, however, is delicate and profoundly affected by altered hydrological conditions. Fen drainage destroys the complex assemblages of marsh-like vegetation resulting from the mixing of acid peaty soils with alkaline ground water. These conditions support a narrow range of specialised species and at Woodwalton Fen over 400 such species are found including the nationally rare fen violet and the Deptford pink.

Mixed fen is dominated by tussock-forming grasses such as purple quailreed, reed canary grass, purple moorgrass, and by rushes such as the blunt flowered rush, black bog rush and by various species of sedge. Meadowsweet, yellow iris, water mint and bulrushes grow within a managed environment which without grazing would degenerate into mixed 'carr'. This in turn is relatively dry and contains a variety of shrubs and trees with a relatively impoverished floristic environment. Reed fen develops from swamps managed to crop the common or Norfolk reed. The great saw sedge develops where the water is stagnant. It produces a dense surface mat of dead litter which resists tree invasion but this gradually changes if left unmanaged and invasion by creeping willow and other trees occurs. Drainage, however, eliminates the standing water essential to reed growth and ultimately produces open pasture or the potential for arable cultivation given sufficiently lowered and controlled water tables.

The potential impact of drainage on flora – and other wildlife – is illustrated by the case of Amberley Wild Brooks, Sussex, for which a drainage proposal by the Southern Water Authority was rejected in 1978 at the first public inquiry held under the Land Drainage Act 1976 (Hall 1978, Penning-Rowsell 1978, 1980, 1983a). Over 400 plant species have been recorded within the Brooks which are therefore of major regional significance (Water Space Amenity Commission 1980b). They comprise a large inland freshwater environment with marked internal contrasts in water acidity owing to the juxtaposition of acid peat bogs and chalk substratum producing alkaline stream water. The flora is unusual and probably unique in that it includes all five British duckweeds, all three British species of water millefoil, all five British species of watercress, six out of seven species of British dropworts, 14 out of the 21 British species of pondweeds, 19 species of British sedge and 35 species of British grass.

As with many important wetland ecosystems a diverse and rare flora provides the habitats for a wide range of both common and rare wildlife. One species of fungus dependent on wet conditions was first recorded in the Brooks and another is known only from the Brooks and one locality in France. The area also has a rich and varied population of insects, notably 17 species of dragonfly. Five of the six species of British amphibians breed on the Brooks and of particular note is the crested newt whose British population has declined by 50 per cent in the past 10 years. The Brooks are also internationally noted for wildfowl which, along with the smaller birds like warblers, would have been eliminated here if drainage had proceeded, as they have been in adjacent drained Brooks.

HABITAT LOSS: THE IMPACT ON BIRDS

Many bird species are dependent upon wetland conditions for breeding, feeding, roosting and refuge. The types of species likely to be affected by drainage obviously depends on the nature of the habitat. The relevant classification of these (Table 5.3) is slightly different from that relevant to the impact of drainage on flora.

Drainage is probably one of the greatest threats to birdlife although drainage is more final in its destruction of plant communities given that birds may disappear temporarily and return if suitable conditions are re-established. As with the impact of drainage on plant life, however, the greatest impact on birds occurs where permanent 'unimproved' wetlands are drained for intensified agricultural use. Here the number of species may be reduced to perhaps to or three breeding pairs per km^2, if any suitable habitats remain, rather than the 20–30 pairs associated with the richest wetland sites (G. J. Thomas, personal communication 1981). The species numbers in neighbouring areas will also be reduced, reflecting the disappearance of the wetland food resources with a lower water table and

Table 5.3 Habitat types and bird species likely to be affected by agricultural drainage (G. J. Thomas, personal communication 1981).

Habitat type	Bird species
narrow 'upland' valleys	dipper, goosander, red breasted merganser, common sandpiper
lowland river environments	moorhen, coot, little grebe, reed warbler, sedge warbler, kingfisher, mallard
lowland marsh habitats	yellow wagtail (many varieties), snipe, great snipe, shoveller, mallard, black-tailed godwits (large marshes)
estuarine and coastal areas	heron, brent (and other) geese, waders, gulls, terns, redshank, knot, dunlin, sanderling, bar-tailed godwits, pintail

diminished surface water flooding.

Birds will not disappear completely of course if an area is drained or reclaimed from the sea, but ordinary 'farmyard' birds will replace the specialist 'wetland' species. For example, to retain their population levels the snipe and godwit require spring overland flooding yet the trends in drainage are to remove summer flooding to maximise the grazing or cropping season. The adult birds need water for food supply or breeding behaviour, as well as the refuge from interference that flooded locations provide. A varied habitat is required, however, so that nesting may take place close to, but not on, the water. If drainage means rapid elimination of surface water the nest that was built near a pool of open water suddenly becomes surrounded by dry land and the birds will have to move or die. For snipe a high water table is important for feeding since it lives on invertebrates taken from below the surface of a wet marsh or bog.

Coastal reclamation such as on the north Kent marshes has reduced the bird habitats drastically (Fig. 5.3). Even conversion of low marsh into high marsh can critically change the conditions for bird life but remain almost unaltered in appearance. Such estuarine and coastal areas are highly productive locations for wildlife by supporting large populations of invertebrates which provide food for wading birds. The Ribble Estuary in Lancashire was the scene of a major drainage proposal in 1978 despite being one of the most important sites in Western Europe for the passage and wintering of wading birds. Breeding birds here include the common tern (about 500 pairs) and redshank (about 240 pairs) and the estuary provides feeding and roosting areas for more than 220 000 waders. These include knot (100 000), dunlin (55 000), sanderling (9 000) and bar-tailed godwit. Several species of wildfowl occur in winter months including up to 10 per cent of the European population of pintail and up to 20 per cent of the world's population of pink-footed geese (Water Space Amenity Commission 1980b). Eventually the proposal came to nothing but only because the Nature Conservancy Council was forced to buy the Estuary to prevent the reclamation.

If coastal reclamation or inland drainage reduces the number of years an area receives surface flooding this affects its use for visiting wildfowl.

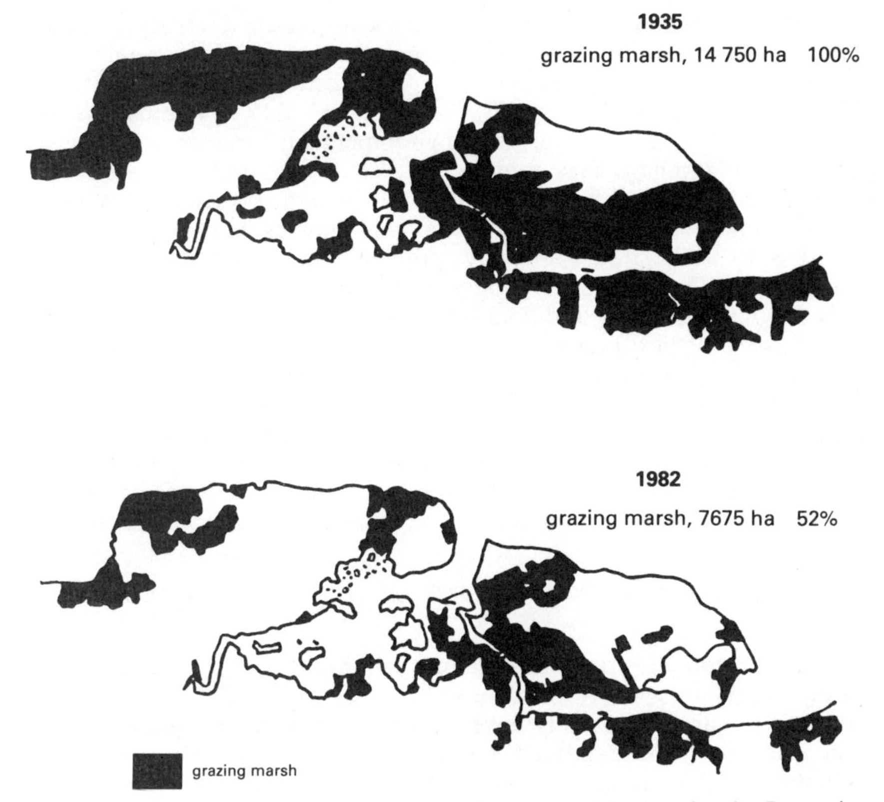

Figure 5.3 Drainage of north Kent marshes (from Royal Society for the Protection of Birds 1983).

Regular flooding such as on the Ouse Washes in Bedfordshire provides a secure roosting site for geese and swans such as the Bewick's Swan but also for the more common mute and the whooper swan which particularly favours saltwater areas (Cadbury 1975). A gradation of flooding depths and durations and of management intensities is required to maintain habitats for a variety of species. For example, cattle grazing in wetland grass areas prevent the build-up of scrub vegetation, which would reduce floristic diversity and so eliminate food supplies, so that the least grazed fields in the Ouse Washes show lowest species count (Thomas *et al.* 1981).

The refuge provided by open water is particularly important at the critical stages in birds' annual cycle such as breeding and moulting when ducks and other passage birds are flightless during high summer and require cover. River regulation for flood control can mean digging trapezoidal channels and removing all overhanging trees. Since the kingfisher requires steep earth banks for its long burrow it cannot then survive. Following decades of river management in the East Anglian Fens the species is now a rarity there (Royal Society for the Protection of Birds 1983).

If drainage reduces the size of wetland habitats this reduces or eliminates their suitability for a number of bird species. Wagtails, snipe and redshank require large breeding territories and are sensitive to disturbance. Apparently small changes from permanent grassland to ley grass rotations can seriously reduce breeding success, as suitable feeding grounds are removed, although birds like the lapwing seem more able to adjust to such disturbance. To preserve the total environment for birds on a 400 hectare site liable to drainage may require at least 20 per cent of that area left undrained and a 0.75–1 km 'buffer zone' between the protected area and the nearest significant human interference (G. J. Thomas, personal communication 1981). Such 'refuges' create problems of access, ownership and management. They may also require to be located in such a way that visiting birds can migrate along traditional flight lines.

The choice of bird preserves to protect from drainage is therefore more complex than that for plant conservation where individual reserves can be viewed a little more in isolation, apart from their regional or national location. Choosing areas to be retained as 'key' bird sites cannot therefore be straightforward, and will depend upon judgement of priorities between sites with a few special species, those with locally significant features, and a site with wide variety of common species. This is by no means easy.

IMPACTS OF DRAINAGE ON OTHER WILDLIFE: AMPHIBIA, MAMMALS AND FISH

Drainage profoundly affects soil–water relationships and as such reduces the variety and density of invertebrate soil life. The invertebrates in reclaimed coastal areas are significantly lower in total numbers and species diversity than when in the pre-drained condition (British Trust for Conservation Volunteers 1976). Amphibians such as frogs, toads and newts depend upon suitable freshwater habitats in which to breed. They depend upon a satisfactory number and spatial density of ponds, pools and ditches – which drainage may reduce or eliminate – and they are also adversely affected by human interference and lack of cover from other predators which agricultural intensification brings.

The only aquatic mammals native to Britain are the otter, water vole and water shrew although the mink and coypu are introduced species which have become naturalised and inhabit fluvial environments despite efforts at control. Otters require relatively unpolluted, undisturbed environments with clumps of vegetation next to watercourses to provide daytime cover. Drainage works along rivers and streams have reduced these environments to such an extent that otters in England and Wales are numbered in hundreds today rather than thousands in the 1950s (Wood 1981). The species is now protected under the Conservation of Wild Creatures and Wild Plants Act 1975 but is still considered 'desperately vulnerable to changes in their habitat' (Royal Society for the Protection of Birds 1983).

Fish require adequate water oxygenation, adequate feed and the protection of their spawning grounds, although different fish species require different levels of oxygen. Fish feed principally on invertebrates that live on or among water plants or which fall into the water from overhanging trees or shrubs. Game fisheries are unlikely to be affected by drainage schemes, since they are both valuable enough themselves to forestall agricultural intensification of adjacent land and are located mainly in upland areas. Although drainage may promote water oxygenation, the clearance of trees and bushes associated with lowland stream and river work may deplete the aquatic organisms that provide fish food and also remove the shade essential on parts of rivers for fish to flourish.

Many types of land drainage and flood alleviation works can also adversely affect fisheries by deepening a shallow stream to increase its capacity, or by removing silt or the gravel river bed in which fish feed, or by removing spawning areas and weed beds. Plant communities in stream and river channels may not regenerate quickly enough following drainage works to provide annual spawning facilities. Weed removal in breeding seasons is detrimental to natural regeneration and the straightening of natural meandering streams and rivers to increase flow rates may eliminate pools which provide protection for fish stocks and the riffles which oxygenate waters (Swales 1982). Only with the most careful site management of river improvement works can fisheries be protected from wholesale destruction (Newbold *et al.* 1983).

THE IMPACT OF FLOOD ALLEVIATION AND DRAINAGE ON RURAL LANDSCAPES

The rivers and streams of Britain constitute one of the principal elements of its cherished landscape. Both major rivers and minor streams contribute substantially to the character and visual interest of lowland areas and form one of the most important stable elements within a rapidly changing and progressively less diverse agricultural scene (Westmacott & Worthington 1974, Shoard 1980). The character of much of Britain's lowland landscape is dependent upon the pattern of hedgerows and woodland groups, many of which exist in conjunction with waterways. The character of the coastal and estuarine marshes depends crucially upon the flat wilderness aspects and the predominance of open water and slow meandering streams.

Table 5.4 An example of the impact of a drainage scheme on landscape features (from Miers 1979).

	Landscape features				
	Trees	Bushes	River shoals	River bends	River cliffs
before the scheme	79	130	21	49	13
after the scheme	33	11	2	21	2
percentage remaining after drainage	41.8	8.5	9.5	42.9	15.4

Many aspects of river engineering, however, can significantly affect these landscapes. The realignment of watercourses may bring artificially regular shapes to what was previously an apparently natural rural scene. The felling of trees associated with drainage works can add to the monotony of already flat landscapes (Table 5.4). The 'rationalisation' of drainage ditches may alter the visual variety of areas once characterised by open water and small-scale fields. Dredging and bank protection may render 'hard' a previously 'soft' and tranquil scene. Buildings such as pumping stations or gauges bring unnatural colours and textures by introducing 'urban' characteristics into rural areas. In totality, drainage may convert unique and historic landscapes into ordinary agricultural scenes.

Sympathetic design can integrate small-scale modern drainage features with traditional landscape character (Water Space Amenity Commission 1980a). It is possible to transform a potential landscape intrusion into a feature that with time will blend into a changing landscape. This can be done by avoiding rigid unnatural shapes, ensuring buildings are sited compatibly with their surroundings, using local or at least native species for planting and by more imaginative use of concrete and other materials. Even major structures, if designed with their visual impact in mind, can present striking features within a flat landscape, just as the windmills of the fen country provide nostalgic reminders of bygone ages and pleasant relief from the boredom of flatness; they too may have been resisted in their day by those concerned to promote an unchanging scene. Nevertheless, where agricultural drainage promotes major changes in land use no amount of careful design can prevent a radical and largely irreversible reduction in total landscape diversity.

Conservation versus drainage: major problem areas

DEGREES OF CONFLICT

The aims and objectives of nature conservation and land drainage are fundamentally in conflict and 'A large part of our wetland heritage has been destroyed or modified by land drainage' (Water Space Amenity Commission 1980a, p. 11). In O'Riordan's (1980d, p. 13) opinion 'The blunt truth is that the method of land drainage and agricultural improvement generally does not take adequately into account the conservation and amenity values. Neither the law nor the deployment of financial resources favours a proper balance between these demands.'

However, the extent of conflict with nature conservation varies between the six main ambits of land drainage (sea defence, reclamation from the sea, defence against fluvial flooding – both urban and agricultural – the lowering of river levels for agriculture and field underdrainage). Of these, land-drainage works for defence against sea flooding generally provide minor conflict because successful completion of a flood mitigation scheme will only serve to maintain the *status quo* rather than result in major newly drained areas. Reclamation from the sea, as in the case of the Ribble estuary or the margins of the Wash, usually causes major conflict between those concerned to preserve existing habitats, principally for birds, and those who want wholesale conversion to relatively barren agricultural areas. Schemes to defend lowland agricultural and urban areas from flooding are most efficient if the engineer can design a straightened channel with clear banks and no aquatic weeds. Such schemes can provoke conflict with amenity interests in leaving stark solutions where tranquil rural streams and rivers once formed significant landscape elements within a pleasant rural scene.

Lowering water tables in agricultural areas, or providing security against flood risk and thereby promoting arable cultivation, will conflict fundamentally with both nature conservation and landscape interests concerned again to maintain the existing diversity of wildlife and scenery. On the other hand, field underdrainage by itself may provoke little controversy in that most is done to support the existing type of farming rather than to promote change; the change comes from the improvements to arterial drainage, which are essential prerequisites to field drainage. The farmer in providing underdrainage is simply reacting to a battle already lost by the conservation interests.

It can thus be seen that the main conflict comes from proposals to drain bogs and marshes for agricultural use, and major schemes to up-grade agricultural productivity and profitability such as in the Waveney Valley or on the Somerset Levels.

DRAINAGE AND THE SOMERSET LEVELS

On the Somerset Levels both continuing peat extraction and increased drainage have already altered extensive areas of wetland habitat. Together they threaten to destroy the rich flora and fauna that contribute to the area's traditional character (Nature Conservancy Council, South West Region 1977).

Drainage on the Levels today is well organised. Maintaining water levels in the rivers and main drainage is the responsibility of Wessex Water Authority which annually spends about £0.75 million on land-drainage improvements and £0.4 million on maintenance (1978). The six Internal Drainage Boards in the area levy rates on the farmers and regularly clean the minor watercourses to maintain flows. Much of the wildlife interest is dependent on the continuation of the traditional predominantly dairying agriculture. This pattern of farming involves restricted cleaning of the drainage ditches, acceptance of the limited winter flooding and a high water table for the greater part of the year.

The Agriculture Advisory Council (1970) indicates, however, that some three-quarters of the Levels are capable of high agricultural productivity with arable cultivation under the right drainage conditions. The Water Authority is continually anticipating demands for improved control of the water table and the capacity of the main river arterial watercourse is sufficient for considerably greater field drainage to lower water table levels throughout the year. Such improvements, with Ministry of Agriculture, Fisheries and Food grant aid, appear cost-effective in terms of tangible economic analysis without consideration of their substantial intangible effects; thus they are proceeding. Peat extraction continues to expand because of increased demand and use of modern technology and owing to difficulties of control over land on which the peat companies have long-standing planning consents.

The Levels may appear barren of wildlife but the open water of the drainage channels can be richly colonised by aquatic plants and animals and there are floristically rich marshlands and wet woodlands. A high water table keeps soil invertebrates near the surface where they provide food for many thousands of migrant and wintering birds. Wading birds and wildfowl number many thousands and the area is the best breeding ground for waders in south-west England. The Levels are rich in insect and other invertebrate life and are considered one of the best localities for the otherwise scarce otter.

Drainage could destroy many of the environments essential for this wildlife and surveys have been undertaken to determine locations of particular conservation value (Fig. 5.4). Limited areas have been set aside for nature reserves, comprising some 141 acres of raised bog and 29 acres of open water and marsh. Nevertheless the continuing trends towards better drainage for intensified grazing and arable cultivation, plus the continued peat extraction, raise serious questions as to the extensive deterioration of the area's wildlife.

What policies can be developed in such conflict situations? Several options have been proposed by the Nature Conservancy Council, South West Region (1977). In essence, these involve either accepting the gradual loss of wildlife or, alternatively, controlling agricultural improvement and peat extraction to maintain wildlife. The latter would involve many interested parties forgoing their rights, aspirations and statutory responsibilities to promote what are seen as economically beneficial agricultural developments by providing improved drainage. A further option is to protect the best areas of the different habitats present – perhaps involving 8 per cent of the Levels – by designating, protecting and managing reserves on a more extensive scale and a more rational basis than those now existing, which have been established as and when land could be purchased or leased by conservation bodies. Such reserves would have to be sufficiently extensive to provide protection of the wide range of species requiring different habitats. They would be particularly expensive to purchase in the case of peat areas.

Such a strategy would require compensation for the farmers thereby not receiving the benefits of drainage if they agreed to continue with their traditional agricultural regimes. A linked and similarly expensive proposal is for the improvement and management of the worked-out peat deposits as open water sites for wildlife conservation. An ambitious land-use strategy for the Levels is proposed in which government departments would agree to the defined use of different areas of the Levels for intensified agriculture, forestry, or conservation. This would involve a 'patchwork quilt' pattern of land use controlled by a clear policy and a form of planning control over a majority of rural land uses. Somerset County Council has produced a local plan based on such a land-use strategy (Water Space Amenity Commission 1980b, Somerset County Council 1983). The implementation of this plan rests with a Countryside Forum as a focus for consultants and with a voluntary notification arrangement for proposed land-use changes. However, it is obvious that this plan is a fragile compromise between interested parties and that the wildlife of the Levels must deteriorate as agricultural use intensifies.

ASSESSMENT OF VISUAL IMPACT OF DRAINAGE IN THE YARE BASIN

Landscape alteration is an emotive issue (Shoard 1980) and no agreement is

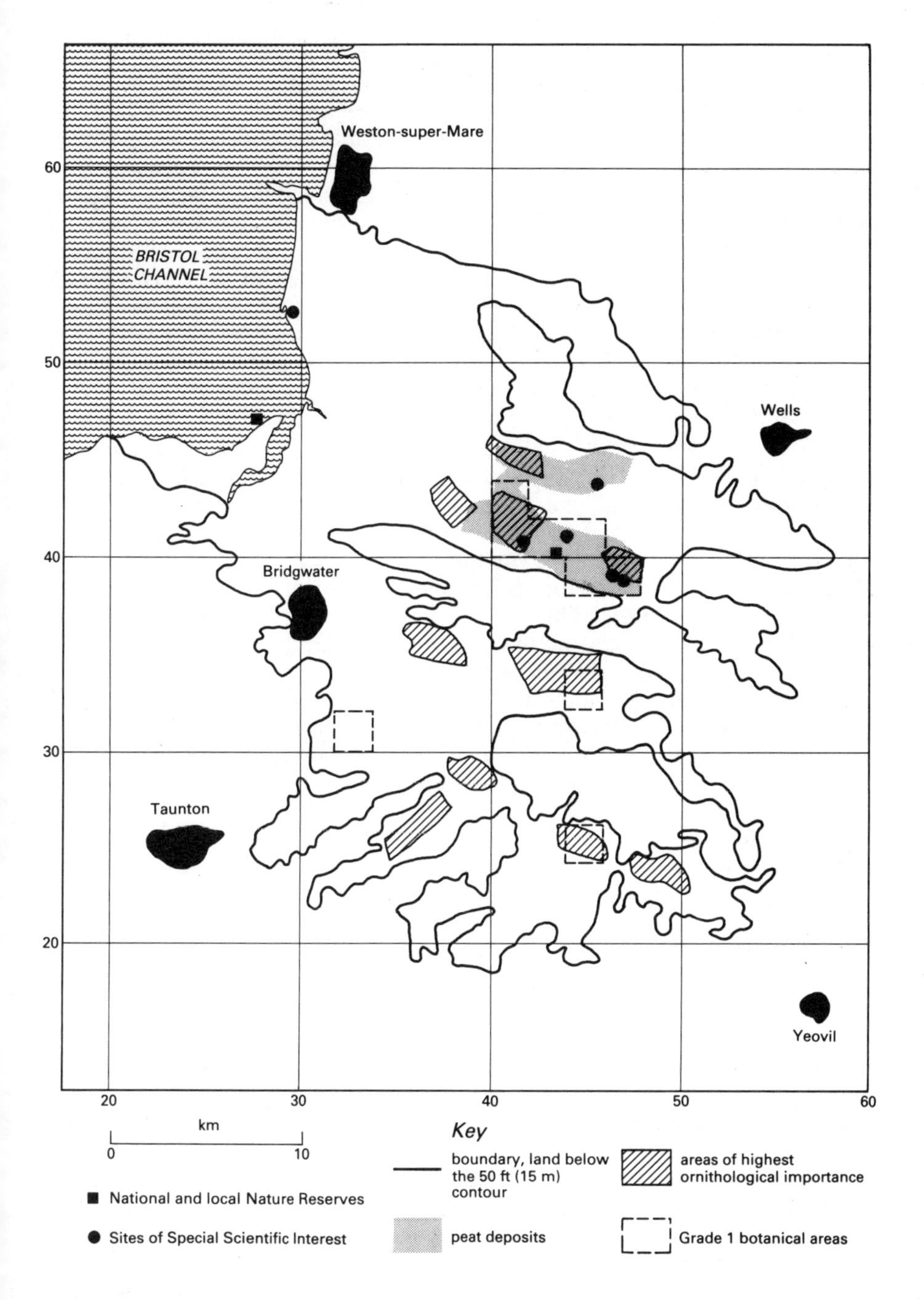

Figure 5.4 Nature reserves, peat extraction areas and Sites of Special Scientific Interest (SSSIs) on the Somerset Levels (from Nature Conservancy Council, South West Region 1977).

likely on ways to assess the qualities of landscapes modified by land drainage (Penning-Rowsell 1981c, 1983a). Nevertheless the arguments against proposals to provide a flood control scheme for the Yare Basin centre around its landscape impact and the secondary effects of drainage on wetland flora. The area is not significantly important for bird life because earlier drainage has converted much of the open water and wetland habitats into summer grazing (George 1977). The area does support significant communities of other animals, higher plants and 18 groups of aquatic invertebrates, many of which are particularly sensitive to habitat disturbance, but the landscape issue is dominant.

The landscape conservation argument (O'Riordan 1980a) is based on the supposition that what is characteristic and a little nostalgic is much more attractive and inspirational than what is monotonous and common to much of lowland England. The lower Yare marshes are vast, so that the dotting of the landscape with trees, hedges, cattle gates and grazing stock gives a pleasant and intimate air to a scene largely unchanged for 200 years. As O'Riordan (1980a) points out, in scale and variety the landscapes are unique to Britain but widespread arable conversion resulting from the more secure drainage following barrier construction would have major visual impact. The landscape would then mirror the intensively cultivated fens of Cambridgeshire and Lincolnshire. The shift from traditional agricultural practices would cause the removal of the many small-scale landscape features that lend so much charm to an historic scene.

The fundamental questions are whether and how to quantify the potential landscape loss to weigh against the potential agricultural gain. As with the wildlife surveys for the Somerset Levels, such landscape characteristics are not impossible to identify and quantify (Penning-Rowsell 1974, 1981c). Research by Land Use Consultants (1978) assessed the landscape resources for the 2800 hectares that comprise the Yare Basin (Penning-Rowsell 1983a). Six major groups of significant landscape elements were identified (landform, vegetation, reed beds, open marshes, animals and human artefacts). Major landscape zones were identified where these characteristics are 'dominant' or 'very significant' landscape features (Fig. 5.5). The assessment is notable in recognising the importance of human artefacts in the landscape's character. The results show an apparent consensus across a wide range of information that the existing features of the landscape are highly prized and that agricultural intensification would have an adverse effect on the visual and cultural attributes of the area.

Such landscape assessments are not without considerable conceptual and technical problems. For example, Land Use Consultants weight all the new elements in the landscape following the possible construction of the Barrier with high negative values which indicate adverse landscape change. This is obviously subjective, and involves value judgements with which not everyone would agree, yet this aspect of the method profoundly influences the results. Furthermore, just as the number and diversity of wildlife species cannot unambiguously characterise or be used to value the totality of a habitat, neither can the sum of individual landscape elements encapsulate the distinctiveness of a scene. Nevertheless some type of landscape classification may at least give some information on the characteristics of valued areas and thus the significance of possible loss of different landscape and cultural features. Indeed, the Broads Authority adopted in 1983 its own landscape assessment as a basis for demarcating drainage 'no-go' areas (Clark 1982).

Environmental impact minimisation

PRINCIPLES OF CONSERVATION AND MANAGEMENT

When seeking to protect plants and wildlife from the adverse effects of drainage the fundamental principles of conservation and management are in essence similar whether protecting otters, reed beds, birds or fish. However, it is also crucial to understand the basic conflict of interest between drainage and wildlife conservation. This understanding clarifies the mind and poses supplementary questions concerning the necessary extent of drainage works and the scale of any potential consequent nature conservation loss.

The first principle is that only active conservation and management of many wetland sites can maintain their wildlife and landscape values. The hydrological regimes of many wetlands are not natural and the existing, often artificial, water table level has to be maintained to prevent successional changes. Creative development of new wetland sites can retain species diversity within land-drainage schemes if funds are available and an appropriate design and management programme is pursued (Fig. 5.6). Grazing is usually essential to preserving the qualities of wet grassland or coastal marshes to prevent scrub or tree regeneration, just as meticulous maintenance is essential to the continued hydraulic effectiveness of drainage works. Reed harvesting is essential to maintain reed fen, and river channels require occasional and sympathetic maintenance to prevent undergrowth choking habitats and reducing wildlife food sources. Only coastal and estuarine marshes require no management.

A second principle of conservation and management is that protecting the totality of the existing environment is almost invariably essential for the maintenance of existing wildlife diversity. Individual components of wetland areas cannot simply be retained in the hope that plant or bird species that require such features will survive; the underlying soil regime or food supply may require different conditions and these may be lost even if the apparently important superficial characteristics are conserved. Attempts to use this principle to separate and protect completely the key areas within drainage schemes from the effects of drainage may be partially successful, but large areas will be required and even then the edges of such reserves may suffer irreversible change owing to influence from their surroundings (Shaw 1979).

The third principle is that adequate research and surveys are an essential prerequisite to any policies aimed at minimising wildlife loss for any given flood protection or drainage scheme. A notable feature of the recent land-drainage controversies has been the stimulus given to surveys and research by government and voluntary bodies (Ratcliffe 1977, Scott 1980, Royal Society for the Protection of Birds 1983, Newbold *et al*. 1983).

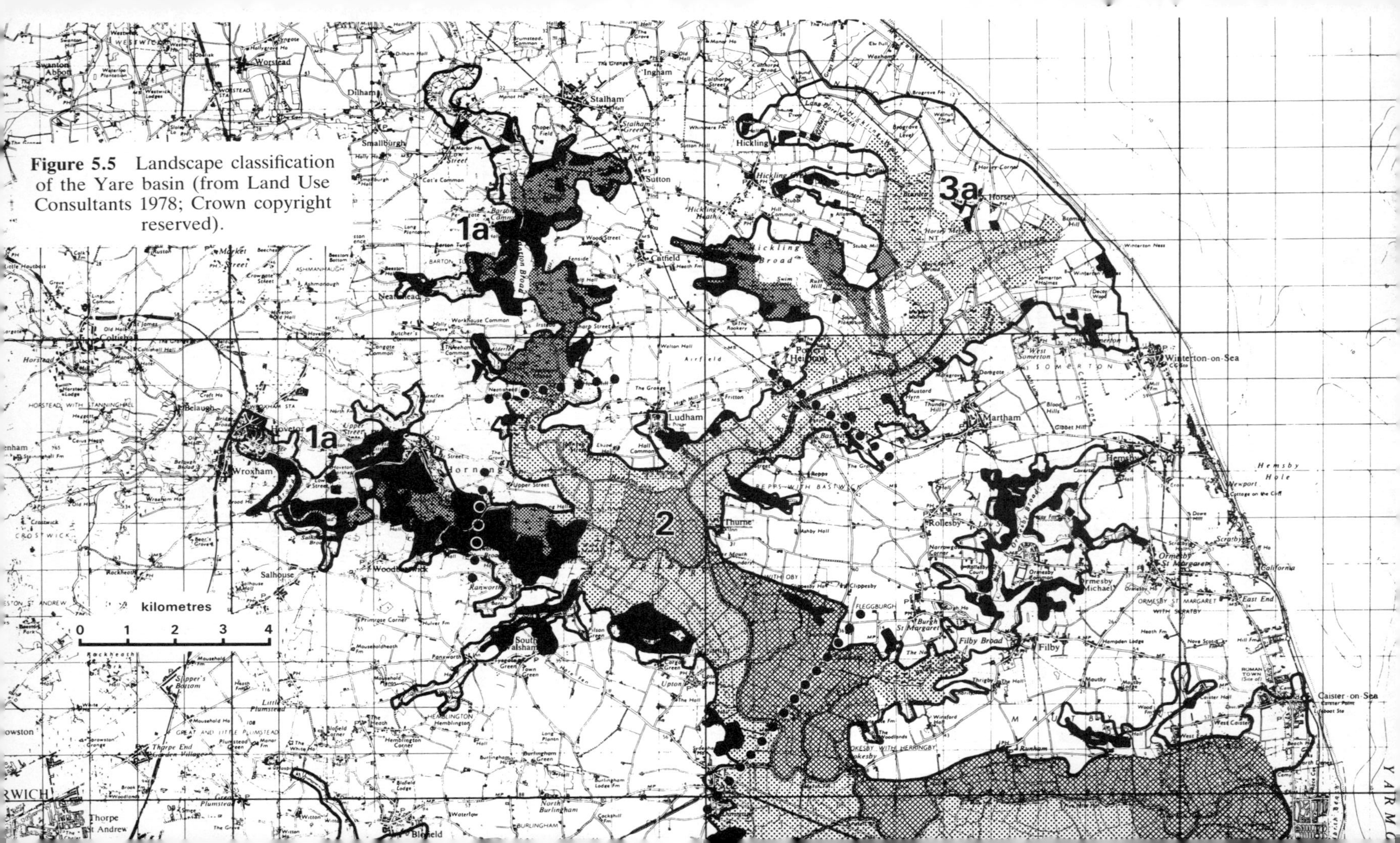

Figure 5.5 Landscape classification of the Yare basin (from Land Use Consultants 1978; Crown copyright reserved).

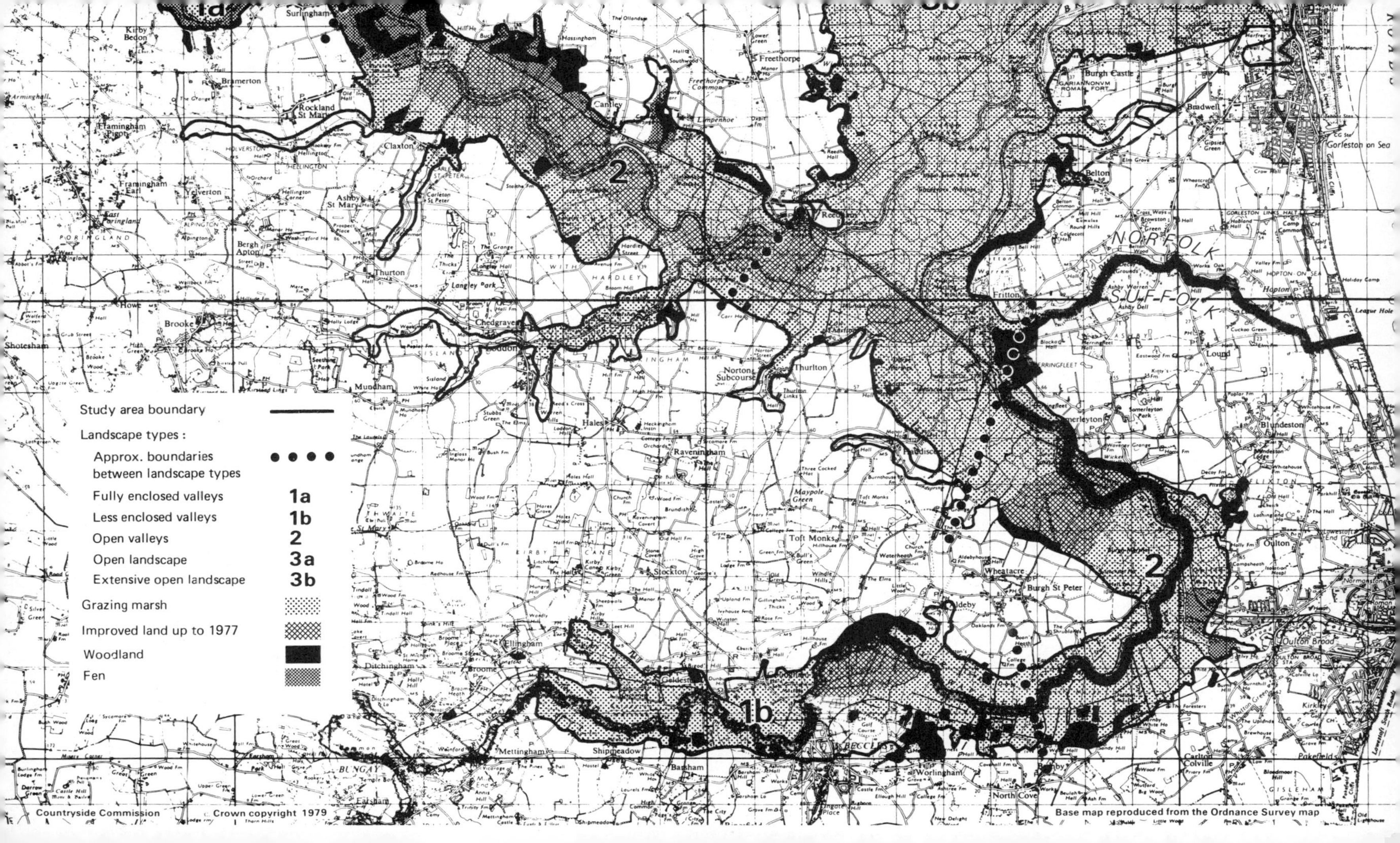
Study area boundary
Landscape types :
Approx. boundaries between landscape types
Fully enclosed valleys 1a
Less enclosed valleys 1b
Open valleys 2
Open landscape 3a
Extensive open landscape 3b
Grazing marsh
Improved land up to 1977
Woodland
Fen
1a
2
2
1b
NORFOLK
SUFFOLK
Surlingham
Bramerton
Rockland St Mary
Claxton
Cantley
Freethorpe
Burgh Castle
Bradwell
Gorleston on Sea
Belton
Fritton
Hopton
Lound
Blundeston
Somerleyton
Oulton
Oulton Broad
Thurlton
Norton Subcourse
Chedgrave
Loddon
Hales
Raveningham
Toft Monks
Wheatacre
Burgh St Peter
Aldeby
Stockton
Ellingham
Ditchingham
Geldeston
Mettingham
Shipmeadow
Barsham
Worlingham
North Cove
Carlton Colville
Bungay
Brooke
Mundham
Thurton
Countryside Commission

Base map reproduced from the Ordnance Survey map

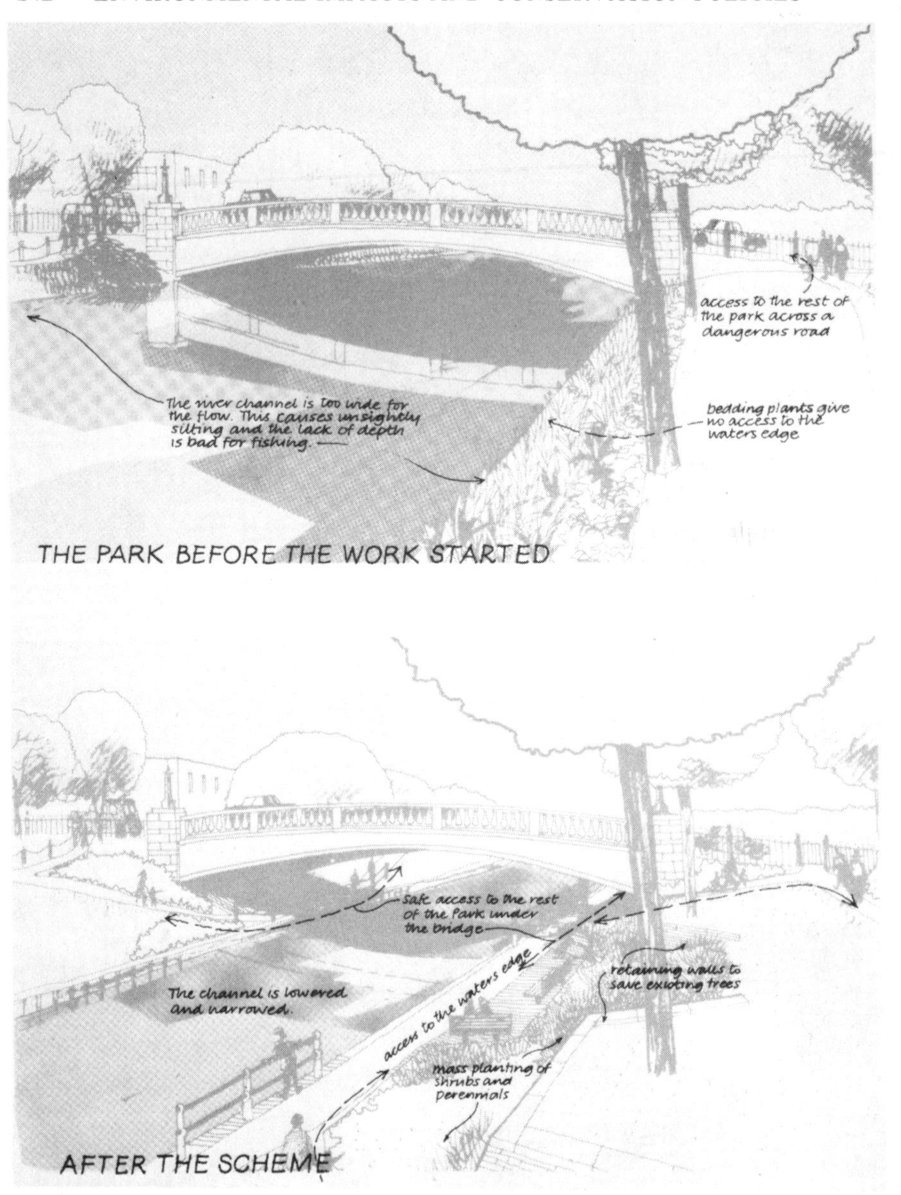

Figure 5.6 Good and bad design of river drainage-improvement works (from Water Space Amenity Commission 1980a).

Surveys can gauge the nature, extent, rarity and hence the significance of plant and wildlife communities in wetland sites (e.g. Ratcliffe & Hattey 1982). Research using longitudinal studies of the effects of drainage in diverse circumstances or comparative studies of drained and undrained environments can be used to gauge the magnitude of potential wildlife loss from different types of schemes. The Royal Society for the Protection of

Birds (1983) and the British Trust for Ornithology are thus embarking upon a major survey of wet grassland sites to determine those sites with rare species or particularly rich bird life. The research results necessary to predict the likely loss of wildlife given varying standards of drainage are sparse, yet only if the individual requirements of each species are known are such predictions possible. Hydrological surveys are essential in identifying the sources of water within wetland sites and thereby the exact environmental conditions that wildlife requires for survival; some sources may be more important than others and these might be retained within any drainage scheme (Gilman & Newson n.d.).

THE IMPLICATIONS

Applying the conservation and management principles must lead to the conclusion that no compromise is possible between drainage and nature conservation in the case of lowland and coastal marshes. To drain these eliminates the very conditions upon which the bird, plant and animal species depend. Birds need the large breeding grounds and aquatic food sources, and the marsh plant species cannot survive without the high water tables and surface flooding that drainage seeks to remove.

In the case of river channels, however, sympathetic management and maintenance can minimise the adverse effects of drainage work. River clearance from just one bank can retain cover, breeding and feeding habitats on the opposite bank while retaining floristic diversity (Newbold *et al.* 1983). Nevertheless the timing of such work is crucial and should not encroach upon breeding sites and spawning grounds at critical seasons. Plants introduced as part of a drainage improvement scheme should accord with the location's characteristics and not be such as to multiply to the detriment of existing species. Compound channels will retain shallows for aquatic plants and maintain deep water at low flows to retain fish communities. Maintenance, however, should be restricted to the minimum and chemical herbicides should always be avoided. Applying many other such principles of maintenance can relatively easily retain or add to wildlife in circumstances where some compromise is possible between the objectives of efficient drainage and the preservation of conservation values (e.g. Paynting 1982).

The Water Space Amenity Commission's (1980a) Conservation and Land Drainage Guidelines propose better design practices for all areas where land drainage is currently in progress. Where a drainage or flood alleviation proposal would have a major environmental impact a comprehensive environmental impact assessment is recommended. In other cases the Guidelines present 'Practice Notes' recommending careful consideration of the specific drainage objectives of the proposed scheme in relation to its environmental context.

However, these Guidelines say little about setting priorities – or resolving fundamental conflicts – but simply designate all wetland as a 'good thing' to be conserved irrespective of relative merit: many problems, therefore, are left untackled. Only the very briefest assessment is made of the adverse visual and ecological impact that unsympathetic schemes for flood alleviation and sea defence can have upon urban environments. Few principles are advocated and little detailed design guidance is given here, perhaps reflecting the essentially 'rural' interests of the Working Party concerned.

It is important to appreciate, however, that these Guidelines were produced within the existing framework of legislation and before the passing of the Wildlife and Countryside Act 1981. More recent recommendations from the Nature Conservancy Council are tougher (Newbold *et al.* 1983), and the Guidelines were revised in 1984, but this still does not alter the fundamental power and responsibility for the decisions concerning drainage schemes. These continue to rest with the local farmer, the Water Authority, the Internal Drainage Board or the local District Council as drainage authority and not, for example, with the Ministry of Agriculture, Fisheries and Food or the Department of the Environment. As such all these guidelines and recommendations seek a compromise between drainage authorities with their duties or powers and the publicly expressed and government supported need to protect wildlife and wetland landscapes. Only in the case of an SSSI is there a presumption against major drainage works, and even then only from the Nature Conservancy Council (Newbold *et al.* 1983), not the original Guidelines (Water Space Amenity Commission 1980a).

What options remain, given that consultation simply cannot resolve fundamental conflict? One 'extreme' solution involves bringing wetlands and associated land under land-use planning control, as advocated by Shoard (1980) for all agricultural land, so that the changes of land use subsequent to drainage would be subject to the democratic control of the local authority. Such a strategy would involve major problems, not least in that most drainage merely results in intensification of existing land uses, rather than the more easily monitored and controlled change of use. A system of planning control would also do nothing to solve the Broadland problem (pp. 164-7) where the real issue is the declining profitability of farming with the drainage *status quo*. Furthermore, some drainage can simply constitute temporary pumping of surface water. It is inconceivable that legislation could control such activities, although major schemes could certainly be subject to some form of planning control as a more satisfactory method of public involvement than the confrontations of public inquiries.

If wetland drainage is not to be the subject of control through planning law, the only real safeguard for major wetland sites is for conservation bodies and government agencies to purchase these areas and thereby control their management. In no other way can the riparian owners' rights to improve their land be permanently curtailed (Ch. 2). However, both acquisition and subsequent management is very expensive. Management is especially costly for aquatic environments profoundly affected by surrounding conditions, as exemplified by the upkeep of the dyke around Woodwaltham Fen (Duffey 1974). Purchasing the Ribble Estuaries cost the Nature Conservancy Council £1.73 million in 1979 and severely limited its activities in other areas for many years thereafter. Furthermore, purchasing funds are unlikely to be found except for key sites: locally significant sites may well be imperilled if acquisition is the only means to reconcile conflicts between agriculture and drainage. Here the only recourse is for management agreements between farmers and local authorities or nature conservation organisations. Within management agreements, however, farmers may demand compensation for not taking advantage of drainage and the increased return they would receive. Such compensation may be needed *ad infinitum* and put an intolerable strain upon the budgets of conservation organisations such as the Nature Conservancy Council and local Naturalist Trusts.

Consultation between interested parties may help to resolve conflict (Ch. 6). Voluntary or mandatory notification procedures can at least allow conservation bodies to intervene to purchase key sites, negotiate management agreements or petition for a public inquiry under Section 96 of the Land Drainage Act 1976 and thereby ensure full discussion of the impact of drainage before it is too late. However, such notification systems have not been particularly successful on a national scale at preserving wetland habitats (Nature Conservancy Council, South West Region 1977). Moreover, such recommendations on consultation and on the need for compromise may be irrelevant given the irreconcilable objectives of drainage and nature conservation and the virtual impossibility that lasting compromises can be reached in circumstances such as the Somerset Levels and the Yare Basin where radical environmental change is proposed. The fundamental question here is whether to drain or not, rather than questions of detailed design or implementation procedures. Moreover, the policies of the Ministry of Agriculture, Fisheries and Food towards increased drainage and those of the Department of the Environment towards nature conservation have quite simply been contradictory (Moore 1980), and such a situation cannot contribute to conflict resolution 'on the ground'.

Assessment

The main driving forces behind flood hazard alleviation and agricultural drainage are the demands for public safety from flooding and the damage that ensues, and the desire for more profitable and productive private farming. These forces result in the implementation of land drainage schemes which undoubtedly have an adverse effect on wildlife and landscape values. This process is supported by a grant-aid system and a cost–benefit analysis procedure which accounts inadequately for the intangible effects of both floods and the potential loss of nature conservation and landscape amenity values from draining wetland sites (Royal Society for the Protection of Birds 1983). The procedure is in turn supported by powerful agricultural interest groups allied to the Ministry of Agriculture, Fisheries and Food.

Against such forces the pressure for nature conservation are puny (Moore 1980), yet they appear to be having increasing impact. Both the Guidelines produced by the Water Space Amenity Commission (1980a, 1983) and the more recent government recommendations on consultation (Department of the Environment *et al.* 1982) show that the Ministry and drainage authorities are acutely aware of the conflict that exists and the need to find some form of resolution. Attitudes are changing. Nevertheless major wetland drainage continues and those concerned to safeguard Britain's wetland heritage, which has suffered so much from drainage in the past 50 years, are required to fight a strenuous rearguard action. Moreover this action has to take place with inadequate data, resources and legislative support to protect even the sites that the government itself has designated as being of national scientific and landscape importance.

6 *Consultation: power, interests and attitudes*

The analysis of power and consultation processes

Those who have command over information, decisions and resource allocation have power. This chapter is concerned with this power and its possible devolution both down the different levels of government and from government agencies to the public at large. Given that there are different attitudes and interests amongst those involved, and that most decisions have distributional consequences, we are concerned here with the consultations and citizen participation that bargaining over this devolution necessitates, and thus with the reality of democracy in our specialist area of land drainage.

This is not, however, an easy field to investigate and interpret, particularly because in Britain 'the invisible political power of influence through social and political connection' (O'Riordan 1981, p. 232) is far more significant than the publicly documented expressions of pressures on decision making. Decisions are often made privately or even secretly leaving little evidence for the observer of the bargaining or consultation processes involved.

Notwithstanding these difficulties, various methods may be employed to analyse power and its devolution through consultation, although each of these methods suffers serious drawbacks. Political scientists (e.g. Richardson *et al.* 1978) suggest that power can be studied by tracing specific causes of public decision making. The aim is to search out critical events that reveal which individuals and organisations were consulted or wielded power to bring about decisions. On the other hand, some sociologists (e.g. Cotgrove 1982) believe that power and influence can be studied by questioning participants about which individuals and organisations are generally powerful and influential in the decision-making process. Alternatively other social scientists advocate various forms of observation, including participant observation, in which researchers 'sit in' on decision-making sessions to discover how decisions are made (Rossi *et al.* 1982).

The methods adopted here include the case-study approach, although the authors' participation in decisions about particular schemes for flood alleviation, land drainage and sea defence is a source of insight. The main advantage of the case-study approach is that it deals with actual events but it relies heavily upon identifying critical incidents and interpreting clues. It is problematic, however, when studying situations in which decisions are avoided rather than made. Participant observation can lead to biased or narrow interpretations, whereas the questioning of participants is limited by the participants' understanding of their own and others' rôles. It may be misleading for analysing particular decisions in which unique assemblages of actors or circumstances are involved.

Rationales for consultation and public participation

The reasons for consultation and public participation are many and various. In the field of land drainage the substance of such consultations embraces negotiation concerning access to land or regarding compensation, environmental protection agreements, and discussions concerning scheme design, cost and contributions from beneficiaries.

From the standpoint of those with power, efficiency of decision making is a major criterion for such consultations. To change policies or to implement plans almost always requires the support or at least the acquiescence of others. Consultations with these interested parties may thus promote efficient decision making and planning by avoiding wasteful confrontation and by minimising mistakes in design and implementation. For example, liaison between land-drainage engineers and those in other spheres of water planning can promote better drainage schemes by allowing integration with recreation or with river regulation for water supply. In addition, participation may educate the public as to the aims of a scheme or policy and forewarn a decision maker about possible adverse environmental or amenity effects of their plans. Inter-agency consultation may promote co-ordination, thus minimising costs or improving design, and forestall claims for compensation or other unforeseen circumstances.

Those making decisions may also consult others to gain their political support. This can be designed to spread the risks inherent in planning and so reduce the possibility of being held personally responsible for mistakes or being caught with the wrong plans if circumstances change: there is always someone else to blame or with whom to share the blame. As such both inter-agency consultation and public participation may appear as seeking some community gain but in reality can just be part of a process of protecting the positions of individuals or agencies and of legitimating decisions reached by those in power (see Fig. 6.1).

There are, however, 'moral' views on consulting those affected by decisions. For example, some consider that it may do individuals 'good' to participate in decision making. Others see that a free-enterprise society implies perpetual choice between alternatives, and that the public thus needs – and has a right to – information from those posing different ways of spending public money: nothing is gained unless it is asked for, but the choice must be informed. Some see that if the state is concerned about the interests of individuals, as is usually suggested, then individuals should be able to decide on issues with which they are familiar and which affect them. From this view comes calls for devolution of decision making to lower levels of government – as local as possible – and away from powerful and unaccountable bureaucracies: plans constructed collectively will be better understood and more readily implemented.

A parallel anti-elitist view regards the professional domination of planning as dangerous and seeks instead a public 'right' to consultation and a more consensual approach to making decisions about resource allocation. Participation has thus been seen as a safeguard against domination by sectional interests. Powerful interest and pressure groups are thus legitimised as being needed to influence government which otherwise is dominated by an immutable ruling class.

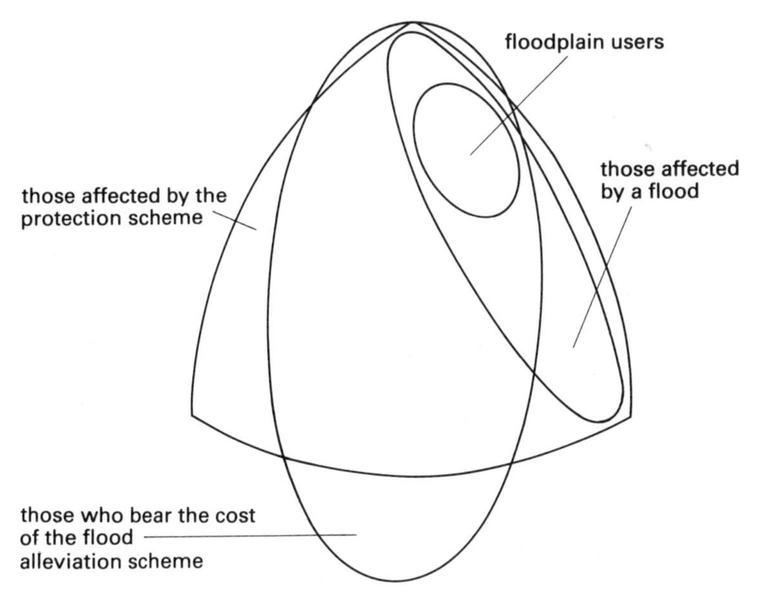

Figure 6.1 The different interests affected by flooding and flood-alleviation schemes (from Green *et al.* 1983b).

There are, of course, counter arguments against consultation and citizen participation in government. Such arguments stress the impracticability of public involvement, especially where the public is shown to be ignorant of technicalities. In the field of flood alleviation such arguments appear to some to be supported by the poor public perception which is common concerning flood problems. Other aspects of similar arguments stress the delay inherent in widening the forum of those involved in decisions: the opposite argument to those seeking efficiency from consultations. Others emphasise the sectional nature of citizen participation pointing to the middle-class domination of amenity societies and pressure groups, which are thus labelled as unrepresentative and therefore to be discounted. Finally those in authority can see devolution and participation as a threat to the basis and structure of their power, and see decisions as better made without the glare of publicity and without exposing the conflict inherent in decisions concerned with spending money.

Degrees and criticisms of consultation

Significantly different levels of consultation and participation exist. At the most basic level those with power can simply proceed to make and implement decisions after notifying others of their intentions. Time and perhaps information is then available for those wishing either to support or oppose land drainage and flood protection schemes. Greater awareness presumably leads to greater involvement in bargaining over policies and implementation. Consultation, in contrast to notification, occurs when interested parties exchange information and views. Compromise can be reached via conciliation when different interests and attitudes occur amongst those involved. External or legal arbitration may be necessary when compromise cannot be secured. Consultation thus clearly requires information to be available to all parties and easy two-way communication and contact between those involved. This in turn requires that resources are available to those wishing to press their views and that representatives of the community and pressure groups are acceptable to those in power.

Full participation, on the other hand, occurs when all interested parties are actively involved in actually making decisions and all take responsibility for these decisions: active co-operation is substituted for conflict and consensus is sought within the decision-making organisation rather than by negotiation between organisations or between agencies and the public.

A CRITICAL PERSPECTIVE

A more critical view of the types of consultation is embodied within Arnstein's (1969) 'ladder of citizen participation'. This hierarchy of consultation types may also be applicable in part to inter-agency consultation and in any case it serves to stimulate our examination of the land drainage consultation processes.

Table 6.1 Arnstein's (1969) ladder of citizen participation.

Rungs on the ladder of citizen participation	Nature of involvement	Degree of power sharing
1 manipulation	rubber-stamp committees	non-participation
2 therapy	powerholders educate or cure citizens	
3 informing	citizens' rights and options are identified	degrees of tokenism
4 consultation	citizens are heard but not necessarily heeded	
5 placation	advice is received from citizens but not acted upon	
6 partnership	trade-offs are negotiated	degrees of citizen power
7 delegated power	citizens are given management power for selected or all parts of programmes	
8 citizen control		

In Arnstein's hierarchy (Table 6.1) non-participation or non-consultation starts with 'manipulation'. This occurs when 'people are placed on rubberstamping committees for the express purpose of "educating" them or engineering their support' (Arnstein 1969, p. 218). These circumstances signify the distortion of participation into 'a public relations vehicle for the powerholders' (p. 218). A more extreme position would be one where interests and attitudes external to those with power are simply ignored. This can be rationalised as the ultimate in representative democracy whereby once citizens have voted for one government or administration no intervention in that system is tolerated: it is up to the government to govern. In contrast Arnstein's 'therapy' level involves those in authority consulting with others so as to 'correct' their misinterpretations and misunderstandings. This coincides with the view that public opinion is generally irrelevant to professional practice and that it should either be ignored or modified.

The hierarchy proceeds with degrees of tokenism: those with power inform others of what they are doing. The emphasis is on a one-way process of notification, with little feedback provided and no power for negotiation. The media and official notices are used for conveying information. With consultation Arnstein sees public hearings and meetings as again giving information to those affected by plans and seeking their views; there is no assurance, however, that public concerns and ideas are taken into account in policy formulation. Placation might involve co-opting members of pressure groups on to decision-making bodies to give advice to those with power who retain, however, the right to judge the utility or legitimacy of such information. Such token consultation means that the influence of the public on decisions is at best marginal; at worst any opposition to those with power becomes internalised within the power groups and therefore diffused.

Only at Arnstein's 'partnership' level is there any power redistribution through negotiation between those with power and those affected by their decisions. Agreement is reached to share planning and decision making through such structures as joint committees and other mechanisms for resolving conflict. A stage further up the hierarchy sees 'citizen power' over one particular aspect of planning, thus assuring the accountability of professionals in that sphere. Citizen control occurs where the public can themselves control both the managerial and policy aspects of planning, by the public's control over the whole programme of works within a planning agency, and are themselves able to negotiate the conditions under which the rules of consultation are drawn up.

This hierarchy can be summarised as beginning with degrees of persuasion based on the idea that people do not know or cannot perceive their real needs and have nothing to contribute to policy making. Tokenism involves degrees of listening to the public or other agencies by those in power, but not doing anything about what is heard. Power control is based on the idea that only citizens have an accurate perception of their needs and the 'right priorities', and that they have the same right to take wrong decisions as all other decision takers.

CONSULTATION IN THE LAND DRAINAGE FIELD: A NORMATIVE VIEW

In the land drainage and flood alleviation field we consider that certain principles should guide the pursuit of adequate consultation and participation mechanisms. We see a need for real devolution, not tokenism, but also a parallel need for education of the public in participation such that their involvement is meaningful rather than based on ignorance or misconceptions. Those proposing change from the *status quo* – whether in the sphere of agricultural drainage or urban flood alleviation – should have the responsibility to consult those affected by their proposals, rather than the onus being upon those affected to react to potential change. Those affected should always be fully informed of potential change to their circumstances in time to react with support or opposition. In addition to thus being informed at an early stage in proceedings, those affected should later be consulted and preferably fully involved in decisions.

This all implies some local forum for consultation and bargaining in which all interested parties are represented and which has some influence over the forces – including the economic forces – promoting drainage and flood alleviation.

Barriers to public consultation and understanding

Liaison and consultation between those with land drainage power and those affected is hampered by institutional complexity, by the perceived remoteness of those in authority, and by the difficulties ordinary people have in understanding the nature of floods and their periodicity.

INSTITUTIONAL BARRIERS

Legal requirements for direct public consultation and involvement in urban drainage proposals are very limited. With local authorities the public has access to policy makers, at least in theory, via their elected local council representatives. Water Authorities are not legally obliged, however, to consult the general public directly over the evaluation, design and implementation of drainage schemes. Matters such as compulsory purchase orders, which may be necessary to undertake works, do require consultation with property owners and, if a drainage proposal is controversial, the Minister may 'call in' the proposal and announce a public inquiry. All aspects of the proposal are then open to public debate and scrutiny although no inland urban flood-alleviation scheme in Britain has been so treated.

Public anger and frustration at inaction by authorities over flooding is not uncommon and it is usually fuelled by the public's poor understanding of the time it takes to plan flood mitigation schemes. Effective communication between authorities and the public is hampered by administrative complexities and by sharp differences in the attitudes and perceptions of the public and engineers. A major cause for concern is that the general public does not understand the complicated division of responsibilities for urban drainage and flood alleviation. This is hardly surprising considering that even some District Councils are confused over their drainage duties! As discussed in Chapter 2 there is also a general failure to appreciate that powers are permissive rather than mandatory.

It is thus easy to understand how public frustration can mount and how accusations of 'buck passing' can arise: different authorities are responsible for apparently similar watercourses in the same locality and the victims of floods find difficulty in tracing those 'responsible' and understanding why months can pass with no apparent action. Ultimately none of the authorities involved is legally obliged to take action to solve a particular flood problem despite members of the public being obliged to pay their local rates and water charges.

Such resentment complements the relatively widespread public perception of Water Authorities as remote and unaccountable bodies responsible for steeply rising water charges: the reputation of local authorities is seldom perceived as better. An attitude of distrust is thus generated and this is exacerbated by an imperfect and indirect form of local representation. Members of the public may have a District or County Council representative on a Water Authority Board or on a regional or local Land Drainage Committee, but few of the public know of or understand these Committees and appreciate their powers. Moreover, these representatives are all appointed rather than elected and for many urban districts there is no appointed representative so that land drainage bodies remain almost totally inaccessible to the public. The reason for this isolation is not conspiratorial but that these organisations are designed as executive agencies to plan land drainage operations, with an emphasis on agricultural drainage, rather than to provide the public with some say in their flood protection.

ATTITUDINAL OBSTACLES TO PUBLIC CONSULTATION

The general public and the engineers charged with solving urban flood problems display significant differences in attitudes and perceptions which concern, first, the flood hazard, and secondly, each other. As first analysed by Kates (1962) the differences in experience between the public and engineers ultimately explain these contrasting attitudes and perceptions.

In contrast to land drainage engineers who work daily with flood problems, a large proportion of the urban public is generally unaware of flood risks to which it is vulnerable (Parker & Harding 1979, Smith & Tobin 1979, Parker & Penning-Rowsell 1982). Public perception of flood hazards is largely a function of flood frequency and residential mobility (Penning-Rowsell 1976). Repeated flooding leads to an accumulation of flooding experience and thereby to flood risk awareness and response (see Fig. 7.1). However, most urban communities have a relatively high residential mobility so that experience of flooding does not accumulate. High residential mobility has become an important factor perpetuating urban flood hazards because periods of residence are brief compared with the return periods of serious floods which are commonly at least 10 or 20 years. Residential mobility reduces aggregate awareness of flood risks because residents with flooding experience are nearly always replaced by those without (Parker & Penning-Rowsell 1982). Because public awareness of flood risk is generally poor, surprise and anger directed towards authorities is the common public reaction to flooding. Immediately following a flood the event is viewed as being totally unacceptable and unquestionably worthy of immediate preventative action almost irrespective of the public cost. This, of course, is anathema to the cost-conscious Water Authority accountant even if it is welcome to the job preservation-orientated land drainage engineer.

However, public concern for flooding is generally remarkably shortlived. Apart from exceptionally serious flood problems where fear may become a daily consequence, complacency amongst the affected public typically takes over after only a few months as memories fade and anger subsides (Parker & Penning-Rowsell 1980, pp. 225–9). In contrast to the land drainage engineers' experience, flooding events are seen by the public as being rare compared with more pressing everyday environmental and social problems such as traffic noise or the threat of unemployment (Penning-Rowsell 1972, Harding & Parker 1974).

Although the risk of flooding may always be present it may not be so perceived: relatively infrequent but serious events are the only form in which the risk is recognised. Serious floods are sometimes remembered as 'one-off' or freak events so that there is no expectation of recurrence. Other residents will inevitably move out of the flood-prone area and those planning a move will no longer concern themselves with the flooding problem (Burton *et al.* 1978). Thus the aggregate effect over time is a loss of awareness and interest in flooding as a problem, and less strenuous effort to seek solutions.

Ignorance about flooding is not surprising since public experience of floods is generally limited. Moreover, unlike the engineer, the public finds difficulty in understanding technical concepts particularly the probabilistic character of natural hazards. Misconceptions are common in press reports and at public meetings following flooding. It is thus relatively common for an angry public wrongly to attribute the cause of flooding and thus misunderstand the land drainage engineers' proposed solutions.

The degree to which members of the public can be expected to think probabilistically must be limited given that their flooding experience may be either slight or non-existent, such that their inferences about the future must be made from ridiculously small samples of experience (Slovic *et al.* 1974). The return period concept is frequently misunderstood. A '30-year flood' is commonly interpreted to be the flood which will occur regularly every 30 years rather than the flood which will occur on average once in 30 years: there is no question of regularity in the engineers' concept of return period but there often is in the public perception. The unfortunate result is that those who have experienced what is announced to have been a 10-year event are confused and angered by a repetition the next year. To them the engineers' 10-year design standard thus becomes discredited and distrust is compounded.

A further problem is the public's perception of flood 'prevention'. There is widespread misunderstanding – apparently extending to local authority planners (Wilkins 1980, p. 127) – of the fact that flood protection works are only effective up to their design standard. The public fails therefore to appreciate that flood banks or sea defences can be overtopped because they are constructed to a finite standard so as not to waste public money protecting against very rare events. Following floods the public tends to demand complete flood protection, not realising that this, in theory at least, is physically impossible as well as economically unviable. Apart from demanding this absolute protection there is no clear public view of desired protection standards, again probably because probabilistic thinking is required. Thus public consultation over design standards is particularly problematical for the engineer and advocacy of standards below the return period of recently experienced events is positively dangerous.

Engineers often believe that the public sees them as cold and uncaring when faced with multiple personal disasters. In reality the engineers witnessing flood damage may be genuinely upset but may well feel frustrated at being shackled by an administrative system that has been, and is, slow to respond by providing sandbags or the finance for the necessary drainage works to prevent what they see. In addition, the public – comprising a variety of groups with different shades of opinion according to their interests affected – rarely speaks with one voice. Thus engineers concerned to consult 'the public' find themselves confronted with what appears to be an incomprehensibly complex and probably irreconcilable set of views.

Given these public attitudes towards flooding, land-drainage engineers perhaps cannot be blamed for being reluctant to consult the public about flooding problems and their solutions. Engineers who work on a series of flood problems in a variety of areas see a depressingly familiar pattern of public attitudes. Urban riparian owners often fail to fulfil their watercourse maintenance responsibilities. Because of limited time horizons and limited experience residents remain unconcerned about flood risks until disaster strikes them and then indifference re-emerges as the memory fades. Time and again property owners are therefore caught unaware of flood risks affecting them.

The public is often apathetic to drainage improvement proposals to anticipate serious flooding and unwilling to see local rates burdened with the cost of apparently unnecessary works. Nevertheless the same people are outraged by flood damage and quick to blame Water Authorities or local authorities, whose responsibilities are misunderstood, for lack of foresight. In this pattern public meetings following flooding commonly take the form of a confrontation. The engineers, unfairly in their view, are harangued for their complacency concerning the flooding problems, which may ultimately be attributable to public carelessness and ignorance, and over which the same engineers may in fact have been struggling to secure solutions for many years.

Current statutory liaison procedures

For certain aspects of flood alleviation and land drainage the law determines liaison and consultation procedures by prescribing who should communicate with whom over what. These statutory requirements are designed primarily to promote co-ordination of the policies and proposals of the many different agencies with interests in urban or agricultural drainage. Such arrangements mainly concern liaison and consultation between statutorily established executive or advisory agencies. They are needed principally because drainage responsibilities are fragmented: without consultation drainage works might be ill-considered and policies could be contradictory or even counterproductive. This liaison primarily concerns the need for the many agencies involved to obtain a 'consent', to 'consult', to provide information or to 'notify' each other about their plans and operations. In addition Water Authorities are required 'to have regard to' certain aspects of particular environmental concern, and under Section 18 of the Water Act 1973 they must establish and consult advisory fisheries committees including over cases where drainage works may adversely affect their interests.

As such, statutory liaison requirements represent the bare bones of a consultation framework. In practice this is considerably extended by a range of informal voluntary liaison and advisory systems both within drainage authorities and between them and other organisations, as explained below.

LIAISON VIA THE CONSENTS PROCEDURE

For any work in, under, over, or adjacent to 'main' river watercourses a Land Drainage Consent must be obtained from the Water Authority acting in its overall supervisory rôle (Ch. 2). This restricted but effective form of notification and consultation is necessary to ensure that land drainage is not impeded by these works, such as bridges or gas lines, and flooding thereby caused (Sections 28, 29 and 34 of the Land Drainage Act 1976). These liaison requirements are designed to ensure consistent standards of drainage works. The aim is to retain strong Water Authority control to promote the overall efficiency of the drainage system, for which many authorities and individual riparian owners have local responsibility, by ensuring that agencies do nothing to reduce the effectiveness of each others' drainage operations. Thus local authorities who wish to undertake land drainage improvements on 'non-main' rivers must also obtain a consent from the Water Authority before proceeding. Other cases of legally prescribed liaison by seeking consent include where one Internal Drainage Board plans to undertake works which will affect the drainage works of another Board: the Water Authority's consent must first be obtained.

TOWN AND COUNTRY PLANNING LIAISON AND CONSULTATION

The law obliges land drainage authorities to consult each other and, for example, Water Authorities are required to consult every local authority in their area about their flood-alleviation and land drainage surveys and plans (Section 24(5) Water Act 1973). Such consultations should ensure that all drainage problems are identified and that available information is properly utilised.

Water Authorities are also required to have regard to any structure plan, local plan or development plan under the Town and Country Planning Act 1971. This makes possible, but by no means ensures, the co-ordination of flood alleviation and land drainage with local authority land-use plans. For example, structure plan forecasts of future urban expansion have implications for storm sewer provisions but the forecasts may in practice be too vague to allow the detailed planning of urban drainage (Penning-Rowsell 1982b).

Local planning authorities must in turn consult Water Authorities about any proposed developments involving works affecting the river bed or banks (Town and Country Planning Act 1971, General Development Order). Under their supervisory rôle Water Authorities also seek to control all development likely to affect drainage adversely, through consultation and liaison with planning authorities. Such consultation procedures are set out in the Department of the Environment's circular 17/82 but the local authority may quite legally choose to set aside a Water Authority's advice against development in flood-prone areas since it is the superior land-use planning authority (Ch. 2).

In line with the provision of information as an important prerequisite for consultation, Water Authorities must send land drainage surveys and plans, including their Section 24(5) land drainage survey reports (Ch. 3), to all local authorities within their areas. These reports must also be available for purchase by members of the public (Section 24 Water Act 1973). The provision of this information on areas liable to flooding should ensure local authority caution in locating new housing or other developments there. It also at least allows the public access to data on the extent of hazards. Prior to 1973 this information was often considered by River Authorities as best left unpublished because the data were liable to error and public release might have created alarm (Penning-Rowsell & Parker 1974).

STATUTORY NOTIFICATION: DRAINAGE AND NATURE CONSERVATION

An important and controversial notification requirement concerns land drainage and conservation (Department of the Environment *et al.* 1982). To promote the consideration of nature conservation Section 22(3) of the Water Act 1973 requires the Nature Conservancy Council to notify Water Authorities of the existence of an SSSI. This notification procedure was expanded in Part II of the Wildlife and Countryside Act 1981 which imposed reciprocal obligations on Water Authorities and other landowners to notify and consult conservation bodies about their drainage plans for any SSSI (Newbold *et al.* 1983, Blenkharn 1983, Water Space Amenity Commission 1983).

As with most notification procedures, however, drainage authorities and landowners are not necessarily obliged to heed the implied warning, and statutory procedures tend only to concern statutory designations (such as an SSSI). Farmers can still proceed to drain important wetlands, including an SSSI, as long as certain rules are followed. For example, a landowner can either gain the written consent of the Nature Conservancy Council to drain, or enter into a management agreement with the Council to manage the land in accordance with conservation interests, or simply wait three months from the date of notification before altering the land or its wildlife.

The ultimate effect of notification is therefore only to gain time for conservation bodies to marshal their arguments for leaving land and wildlife undisturbed. The maximum penalty for not following the statutory rules is a fine of only £500. Clearly, however, these procedures of statutory notification can and do lead to informal bargaining processes, particularly where the Council's principal weapon, the management agreement, is pursued. Under such agreements landowners can be paid for not draining their land. Farmers may thereby be compensated for the profit they might lose by not undertaking agricultural operations which would raise output but destroy wildlife or alter significant landscapes: they can therefore be paid for doing nothing. This question of compensation remains controversial and contrary to the situation in urban areas, where the principle of community land-use control without compensation for development value thereby lost was established in the 1950s. More recently ideas are shifting towards paying the farmers to undertake a conscious conservation 'stewardship' of their valuable wetland areas rather than merely paying for inactivity.

The Wildlife and Countryside Act 1981 does, however, give Ministers certain additional powers which mean that in certain circumstances landowners may be prevented from draining their land. The Nature

Conservancy Council can request the Minister to issue a nature conservation Order. These may be issued where the Council fears that the land will be altered before the negotiations with the owner to leave the land undisturbed have been completed. The first such Order giving special protection was issued during 1982 to a Hampshire landowner who planned to drain a wetland area and to farm it for beef cattle.

This type of Order may forbid landowners to alter their land or remove any flora or fauna. Landowners then have the right to appeal against the Order but if the Order is upheld the Nature Conservancy Council may then acquire the land by compulsory purchase. This is the only recourse if the landowner does not agree to the compensation which must be offered and also refuses to conserve wildlife on the land. Landowners who are convicted of contravening an Order are liable to restore the land to its former condition or pay a fine which increases for each day during which the offence continues.

The entire process associated with the making of Orders, the compulsory purchase of land and the agreement on compensation is likely to lead to protracted debate and bargaining. Individuals and agencies will thus seek to exploit each others' weak positions and public opinion. Pressure groups will seek to influence decision makers and all interests concerned will seek to swing decisions in their favour. The situation is made much more complex by the fact that the Nature Conservancy Council has very limited funds to 'buy out' farmers. The powers are therefore in theory available but in reality many problems remain.

Consultation and bargaining

FORMAL AND INFORMAL PROCESSES

All decisions, whether resulting in new policies, new schemes or even no decisions, have distributional consequences: some people or groups gain; others inevitably lose (Fig. 6.1). Where it takes place, consultation invariably concerns trading in these distributional effects of alternative courses of action. In this context the statutory consultation requirements are not insignificant but consultation affecting decisions takes place both formally and informally. Formal consultations are the most visible expressions of influence and power but informal 'behind the scenes' negotiations, agreements or disagreements are probably much more important (O'Riordan 1981, p. 232).

The formal procedures provide the basic guides or 'rules' for consultation but these may be preceded by, may follow, or may be accompanied by an informal and sometimes a totally private process of bargaining. The degree to which issues grow into more protracted disputes involving extended consultation, bargaining and statutory liaison depends on a number of factors. These include the perceived extent of potential gain or loss on each side and how much the distributional consequences of decisions are clear or become known to those affected. When the bargaining takes place also depends on the degree to which those with power decide to promote, allow or are forced into meaningful consultations. If they are, then these may extend into either formal or informal bargaining and negotiation in order to reach a compromise so that a decision may be taken or implemented.

BARGAINING AND MANIPULATION

Bargaining usually involves a process of public or private 'posturing'. Agencies or interest groups signal what their reactions might be if certain decisions were taken. Proposals may then be modified to meet more nearly the objectives of the various interests involved, perhaps so that some decision is taken rather than a stalemate continue. Concessions may be forced from those with power by 'threats' of protracted public inquiries or by media exposure. Conservation groups may offer concessions from their ideal, such as forgoing certain minor wetland protection, for fear of being ignored completely.

Decision takers may also be deliberately manipulative, seeking to conceal issues or suppress the flow of information to affected parties. This is often done for fear of stimulating public debate and opposition or delay in decision implementation, all of which might adversely affect the official agency or its officers. The manipulative strategy may be successful, although concealing issues and withholding information is often a cause rather than a cure for conflict and delay. Each stage for each actor, however, is a gamble with uncertain odds and often unforeseeable consequences.

Direct public consultations over proposals for urban flood alleviation usually go well beyond the bare legal requirements. However, they still fall short of full public participation and involvement in decision making. A Water Authority may well provide the public with factual information either through a press release, publicity pamphlet or at public meetings. Personal contact between engineers and flood victims or local opinion leaders 'behind the scenes' may also prove important in explaining proposals, and in some cases in modifying amenity aspects. The public meetings may be used to 'sound out' public opinion prior to making decisions, but they are perhaps more frequently used either to reveal limited information about predetermined proposals or, at worst, simply to placate a public angered by floods or by apparent official inaction.

DISPUTES AND CONFLICT

Disputes are not everyday occurrences and some important decisions appear to attract little public attention. The interests affected may be unaware of decisions being made. Alternatively, decisions may appear to be routine and consequence-free, or deliberately made to appear as such. Certain cases or issues may, however, fall below some threshold level of active controversy. In these cases political alignments may be too shaky to create the coalition of interests necessary for concerted political action and to precipitate bargaining. Conservation or other interest groups also cannot 'fight' all cases and all issues: some may be abandoned for lack of resources or local support.

Nevertheless some issues, particularly those in the field of agricultural land drainage, arouse extended and heated political conflict. Such conflict appears to arise in a number of circumstances. For some public issues a rather well defined and enduring coalition of interests emerges so that when 'conflict arenas' appear supporters align themselves quickly and firmly to one side or the other. This may arise where there is a history of conflict between opposing groups – as there is now between farming and conservation interests in England and Wales – and where the distributional consequences of proposals are well recognised and even anticipated.

Conflict surrounding a particular drainage proposal may therefore be just the latest in a long war of attrition between the opposing interests. Those adversely affected by proposals frequently claim that consultation has been inadequate. A multi-stage political tussle ensues in which affected interests seek by various means to halt the opposition and to develop their bargaining power. The means may include litigation, lobbying, attempts at law-changing and recourse to other political and legal mechanisms. At public inquiries or similar 'confrontations' key researchers or academics will be used to exploit their supposed neutrality. Each side will compete for credibility by using the most renowned 'experts' and most senior lawyers they can afford.

The result can be stalemate or complete victory for one side, but if bargaining starts it takes the form of concession-trading in order to adjust distributional consequences to a mutually tolerable balance. Negotiation may take place during a public inquiry or through closed-door arbitration between interested parties. The outcome is usually either some form of compromise, in which agreement is reached over the distribution of benefits and costs of decisions, or it may be that the most powerful interests 'win' at the expense of others, who continue to oppose and dissociate themselves from the decision. Thus the conflict may become further extended over different cases and locations, and the groups or interests fight successive battles in which the weak seek – and may ultimately gain – some form of institutional reform or a change in attitude.

Improving liaison, consultation and co-operation

Although there are undoubtedly many barriers to improved consultation, and certain controversies have proved very protracted, significant attempts towards strengthening liaison and consultation have occurred recently. Two such attempts – one each from the fields of urban drainage and agricultural drainage – are examined below.

INTER-AGENCY LIAISON AND CO-OPERATION: THE CASE OF URBAN WATERCOURSES

With the tangle of existing legislation on land drainage, an effective urban drainage programme can only be secured through co-operation and a genuine willingness to employ permissive powers. A government consultative paper on land drainage powers, published in 1978, referred to the 'confusion regarding responsibilities for maintaining and improving watercourses in urban areas' (Ministry of Agriculture, Fisheries and Food 1978b) but failed to produce policies to clarify and simplify these responsibilities. This confusion of responsibilities between local authorities and water authorities, complicated by the multiplicity of private riparian owners in urban areas, leads to the need for considerable liaison and consultation between those involved.

Water Authorities have had considerable difficulty in persuading small District Councils to improve their small 'non-main' watercourses. The Councils have a clear statutory responsibility for this but, given their powers are permissive, they have been reluctant to undertake the necessary works since they have often lacked sufficient funds. District Councils have argued that Water Authorities should be responsible for all watercourses but Water Authorities retort that countless practical problems would arise from complete Water Authority control over the myriad small local watercourses with flooding problems, not least of which is a statutory definition of a watercourse (Wilkins & Lucas 1980). The failure of many individual riparian owners to fulfil their legal responsibilities to maintain water flow is an added problem. Although riparian owners may be persuaded and can ultimately be forced to undertake river improvement work, this is rarely feasible in practice because of the large numbers of owners involved in urban areas and the co-ordination difficulties that this situation brings.

Complaints that Water Authority and District Council consultations over urban drainage are non-existent or completely inadequate have been frequent (Wilkins & Lucas 1980). Instead of liaison, acrimonious arguments have ensued over responsibilities and the lack of a final obligation on the part of either authority to undertake drainage works. These difficulties may be attributed to legislative deficiencies or to inadequate consultation arrangements based in part upon unco-operative attitudes which have their roots in the different interests of the authorities and individuals concerned. Some District Council representatives believe that law is adequate and workable providing there is a willingness on behalf of everyone concerned to co-operate (Wood 1980), which sometimes there is not.

One attempt to develop co-operation between local authorities and Water Authorities is embraced in the Severn Trent Water Authority's (1977) plan for 'A unified approach to land drainage'. This approach advocated the 'maining' of many 'non-main' rivers, thus progressively extending Water Authority control over urban watercourses, and improving the liaison between all drainage authorities with permissive powers. Through improved liaison Severn Trent hoped to make District and County Councils more aware of the need for drainage improvement, based on priorities set in the Section 24(5) land-drainage surveys, and would thus be persuaded to use their own permissive powers.

Not all local authorities welcomed these proposals, however, some treating them as a further move to reduce the powers of local authorities. Moreover, effective liaison is problematic with the 90 District Councils and 19 County Councils in the Severn Trent Water Authority area, although the Water Authority suggested that District Councils develop more effective liaison links via the County Councils. Extending the main river system upstream of existing main rivers ultimately requires the agreement of each local authority concerned and one missing link in the chain, as happened in some instances, can prevent the overall maining policy from succeeding.

Five years after producing 'A unified approach to land drainage' the Severn Trent Water Authority (1983) considered that this policy, together with a continuing dialogue with local authorities, had resulted in an improved attitude and response by local authorities. The definitive and widely disseminated statement of the Authority's responsibilities had clarified the legal situation where before there had been much confusion and misunderstanding about drainage responsibilities. Between 1977 and 1983 the length of main river increased by 5.3 per cent (from 3275 to 3449 km) and total local authority expenditure reached £5.4 million in 1982–3 and thus just exceeded 50 per cent of the Water Authority's total (£10.2 million).

A DEVELOPING SYSTEM OF VOLUNTARY CONSULTATIONS WITH CONSERVATION INTERESTS

In contrast to the statute-based system for tapping local farming opinion on drainage matters, via their representation on Land Drainage Committees, is the comparative absence of similar arrangements for amenity and nature conservation interests. However, the law concerning liaison between drainage and conservation interests has been complemented by a system of voluntary codes of conduct in the form of the Conservation and Land Drainage Guidelines, to promote the necessary consultations, and some more recent advice from the Department of the Environment *et al.* (1982).

These Guidelines include recommendations on the consultations necessary to reduce conflict between drainage and conservation interests as perceived by a Water Space Amenity Commission (1980a,b) working party, based on discussions with the major statutory and voluntary bodies involved. However, the Guidelines have no legal backing and are advisory only; grant aid can still be paid by the Ministry of Agriculture on a scheme for which the recommendations have been ignored, although this is unlikely. None of the bodies consulted is officially committed to particular proposals or suggestions.

The Guidelines propose that better local consultative arrangements and procedures should be devised and agreed between drainage authorities, local authorities and local conservation groups. An indicative checklist of interests to be consulted is provided (Table 6.2) and several important forms of consultation are recognised. The first concerns 'imperative' consultations: those necessary to avoid legal problems arising from drainage proposals or works (Table 6.3). Early or 'advance' consultations over the environmental implications of drainage proposals are considered crucial. The Guidelines suggest that such consultations should identify conflicts of interests as early as possible to allow time for further investigation and negotiation and, ultimately, to design for minimum environmental impact commensurate with cost-effective and efficient drainage improvements.

Further consultation procedures are suggested during the design and execution of capital drainage works: before detailed design commences consideration should be given to their possible effects on nature conservation, landscape, amenity, fisheries, recreation, agriculture, historic buildings and other special interests. The Guidelines advocate careful analysis of the

Table 6.2 Land drainage and conservation: consultation checklist (from Water Space Amenity Commission 1980a).

	Area	Bodies to be consulted
1	any area in respect of which it is considered that the proposed drainage works would have major environmental impact	County Council, District Council, Nature Conservancy Council, local conservation interests
2	National Park	Park planning authority or committee, District Council, local conservation interests
3	'Area of Outstanding Natural Beauty' designated under the National Parks and Access to the Countryside Act 1949	County Council, District Council, local conservation interests
4	(i) 'Heritage Coast' in a National Park (ii) 'Heritage Coast' not in a National Park	as in '2' above as in '3' above
5	areas of great landscape, scientific or historic value designated in local planning authority development plans	as in '3' above
6	'Sites of Special Scientific Interest' designated under the National Parks and access to the Countryside Act 1949	Nature Conservancy Council
7	National Nature Reserve	Nature Conservancy Council
8	other nature reserves	owner or manager
9	Bird Sanctuary as designated under Protection of Birds Acts 1954–76	as in '3' above
10	Conservation Area designated by a local planning authority	County Council, District Council
11	listed building of historical or architectural interest	as in '3' above
12	ancient monument	as '3' above, together with the Director of Ancient Monuments
13	footpaths and other public rights of access to areas of mountain, moor, heath, down, cliffs or foreshore and other places of natural beauty: (i) in a National Park (ii) not in a National Park	 as '2' above as '3' above
14	any urban area	County Council, District Council
15	sea defence area	local sea fisheries committee

hydraulic efficiency of the proposed drainage works so as to explore various compromises between this and the loss of conservation values.

Initial consultations are necessary to identify works likely to have a major environmental impact. A comprehensive survey of environmental impact is

Table 6.3 'Imperative' consultations concerning drainage proposals (from Water Space Amenity Commission 1980a).

It is always imperative:

1 to consult the owner, occupier and manager of the land or water, and any other person known to have any particular legal interest in it

2 to consult the Water Authority, with regard to the interests of river management, including land drainage, fisheries (including commercial fisheries), recreation and amenities

3 to consult the local Internal Drainage Board if its interests in a watercourse or sea defences are concerned

4 to consult the Department of Trade as to any proposed drainage works on or below the foreshore

recommended for these works, to be undertaken by those proposing the scheme in co-operation with other affected interests. Through such a survey, the Guidelines suggest, policies or designs can be identified which will minimise conflicts of interest. Subsequent drainage work would then follow the Guidelines' 'practice notes' (Ch. 5). For works having a minor environmental impact the 'practice notes' should again be followed in addition to any other measures indicated from local consultations. The Guidelines recommend consultation on an annual basis for maintenance works which can have a significant environmental impact.

In practice the Guidelines in part reflected rather than created a considerable increase in the extent of discussions between drainage authorities, farmers and conservationists (Ministry of Agriculture, Fisheries and Food 1981b, pp. 2–3). Personal contacts – perhaps the product of previous confrontations – have strengthened the web of communications between various interests where much depends upon individuals. Following several set-piece 'battles', notably over the proposed drainage of Amberley Wild Brooks (Parker & Penning-Rowsell 1980, pp. 229–34) most Water Authorities now have arrangements for regular consultations at a variety of levels. However, the Guidelines set the scene for the changes in consultation requirements made in the Wildlife and Countryside Act 1981. These in turn led to the consolidation in 1982 of central government guidelines on land drainage and conservation (Department of the Environment *et al.* 1982) and the firm establishment then of widespread consultation within the planning process.

The Guidelines' importance was therefore more subtle than their content suggests in constituting one step in the process of changing professional attitudes and expectations. As such their impact will arguably be more permanent than implied by the mechanistic set of consultation and design specifications they comprise.

Protracted controversies

THE YARE BASIN SAGA

The case of the Yare Basin in the Norfolk and Suffolk Broadlands well demonstrates the deep conflict of interest characterised by extensive bargaining and frequent deadlock which arises from agricultural drainage proposals and their environmental impact. The controversy has centred on the Halvergate Marshes issue and emphasises the inadequacy of present administrative and legal arrangements for resolving such conflicts and the powerful economic forces that lie behind proposals to drain land for agricultural improvement (O'Riordan 1980a,b).

In one of the latest phases of the controversy the Norfolk and Suffolk Local Land Drainage Committee proposed, through the Anglian Water Authority, to build a barrier across the Yare. The aim is to reduce the threat of tidal surge flooding as occurred in 1953 and 1978, and as exacerbated by the secular rise in east coast sea levels. No-one contests that the Yare flooding threat is severe and that new flood defences are needed even to maintain the current level of protection. Without further conversion to arable cultivation based on drainage, farming will become increasingly unprofitable owing to the poor returns from low-intensity grazing. Thus the *status quo* cannot be sustained. Both the flooding and the farming will get worse unless something is done, yet both flood prevention and intensified farming will indirectly or directly destroy the essential character and remarkable ecological value of the area. An agreed compromise is therefore required between the extent of drainage and the preservation of areas of ecological and landscape significance (Caufield 1981).

The powerful economic forces that encourage agricultural intensification mean that drainage is proceeding even without the barrier. A factor here is that many of the pumps used to drain the lower Broads river valleys are becoming obsolete but the Internal Drainage Boards are only eligible for 50 per cent grant aid for modernisation when they can show real agricultural improvements from the investment: a benefit–cost ratio above unity. This encourages the Boards to invest in new high-capacity pumps rather than straight replacement. These enable water tables to be lowered further, thus encouraging landowners to switch from permanent grazing to improved grass and arable cropping. The overall effect of the Ministry of Agriculture, Fisheries and Food policy, and the high prices for cereal production under the Common Agricultural Policy, is to provide a strong incentive for farmers to invest in deep drainage and to convert their land to arable cropping. O'Riordan (1980a) interprets this as public funds being invested in the capital-intensive private sector at the expense of certain environmental values.

Given the complex institutional arrangements for land drainage and land-use planning in Broadland, the Broads Authority has been specially created as the principal planning body for the area. It embraces all the important land drainage, economic development and conservation interests (O'Riordan 1979). The Authority recognises that farming profitability must be safeguarded and wishes to adopt a land-use strategy embodying management agreements to protect the Broads' most important habitats and landscapes while allowing drainage elsewhere.

This form of compromise is expensive, however, in that compensation for not draining land would be based on the cereal gross margins inflated by intervention prices and could be substantial. Furthermore such agreements are in many cases difficult to obtain and may only be temporary, given the highly controversial nature of the compensation arrangements. Some conservationists consider that such compromise is impossible, and farmers are unwilling to give absolute safeguarding guarantees. The Broads Authority has demanded deferment of the barrier decisions until further ecological and landscape investigations are complete. The Authority's opposition to the continuing grant-aided drainage can only be effected, however, on an *ad hoc* basis. This involves the Authority using its delaying tactics and powers of persuasion to force consultations, management agreements or the maintenance of the *status quo*, which in any case is acknowledged to be unsatisfactory.

The Halvergate issue demonstrates in detail the complex process of informal bargaining and manoeuvring towards some compromise between agricultural drainage and environmental interests. The process began in October 1980 when the Bure Marshes Internal Drainage Board notified the Nature Conservancy Council of its proposals to drain land in the Halvergate Marshes – a proposed SSSI. The Marshes are considered the last remaining open marsh landscapes of their type in the UK. During 1980, and independently from the Halvergate proposals, the Broads Authority made an informal agreement with the Ministry of Agriculture, Fisheries and Food that the Broads Authority should find any proposed drainage scheme 'acceptable' before Ministry grant aid is paid. Effectively this allows the Authority to delay drainage proposals and subject such schemes to a form of environmental impact assessment.

In January 1981 the Broads Authority officers recommended that the Halvergate proposals were 'unacceptable' and that if the Ministry were to provide grant aid then a public inquiry should be sought, partly as a threat to force farming interests to compromise. Fearing confrontation and the loss of co-operative relations with farmers, the members of the Authority pursued instead a compromise solution designed to minimise the alteration of landscape character. By February 1981, however, the Authority was widely condemned by local environmental organisations for not accepting its officers' recommendations.

Nevertheless, after this setback a compromise solution was still sought through a series of meetings between the Ministry, the Countryside Commission, the National Farmers' Union, the Nature Conservancy Council, the Country Landowners' Association and the Broads Authority. The compromise was designed to safeguard blocks of marsh landscape but allow drainage elsewhere. Although the Authority originally used the threat of a public inquiry and confrontation, the subsequent search for a compromise meant this weapon came to appear unreal and was abandoned. A crucial issue preventing a compromise was the source of funds for management agreements. Although the drainage interest resolutely refused to contribute, the Broads Authority and Countryside Commission eventually agreed to take the lead by contributing £25 000 thus effectively agreeing to pay for landscape protection.

Although the funding of management agreements remained a barrier to progress a firm compromise had emerged by June 1981 to safeguard two landscape areas (450 ha in total). The Countryside Commission, dissatisfied with this compromise, then decided to call for a public inquiry but was not supported by the Authority. By August 1981 the compromise was in danger of collapse due to continued lack of agreement over funding arrangements for management agreements. Although the Ministry and Department of the Environment produced a compromise with a total £65 000 for compensation, this proved unacceptable to the Broads Authority because of a disagreement over indexing the compensation payments to inflation.

By January 1982, therefore, a stalemate had been reached with drainage and environmental interests in deadlock. During 1982 part of the Halvergate Marshes were ploughed up by a farmer in defiance of environmental interests. However, in November the Ministry used its new powers under the Wildlife and Countryside Act 1981 to protect conservation interests by rejecting applications for grant aid to drain the Marshes thus resolving the key issue.

The case illustrates the problems of reconciling positions which, through a history of conflict, have become entrenched and are in any case diametrically opposed. Entrenchment derives partly from the progressive impact of the powerful economic forces driving the farmers to 'improve' land and the 'last ditch' position perceived by conservation interests who see wetlands becoming increasingly scarce. Entrenchment is, however, also due to the relentless progress of the chess-like political manoeuvring and infighting which steadily antagonise all those concerned. Rather than opt for formal negotiations the Broads Authority members believed that private discussions and bargaining would be more productive in reaching a compromise than a public inquiry, which was seen as divisive and expensive.

As an institutional innovation the Broads Authority was formed ultimately to reconcile these interests, but this has proved impossible. However, the Halvergate saga was eventually settled, perhaps temporarily, by allowing the Ministry to use new powers thus demonstrating a case for institutional change where prolonged deadlock occurs. What forced the Ministry eventually to adopt an uncharacteristic pro-conservation stance was political pressure through Parliament and the media, which threatened to make farming interests appear too extreme, powerful and privileged. This threat was seen by those with influence as potentially leading to greater harm to farming interests than the loss of a particular battle over a specific drainage scheme. The conservationists' 'victory' in fact leaves the factors encouraging drainage unaltered and the political power structure virtually unchanged.

THE WHITSTABLE SEA DEFENCE/COAST PROTECTION CONTROVERSY

Some of the factors further complicating consultations are exemplified by the controversy over Canterbury City Council's proposals for an £11.75 million coast-protection scheme for Whitstable on the north Kent coast (Fig. 6.2).

About one-third of Whitstable's 16 000 population live in an area with some risk of sea flooding. Also flood-prone is the entire retail and commercial centre. Whitstable has a long history of marine encroachment and sea-defence works and, in common with most of England's east coast,

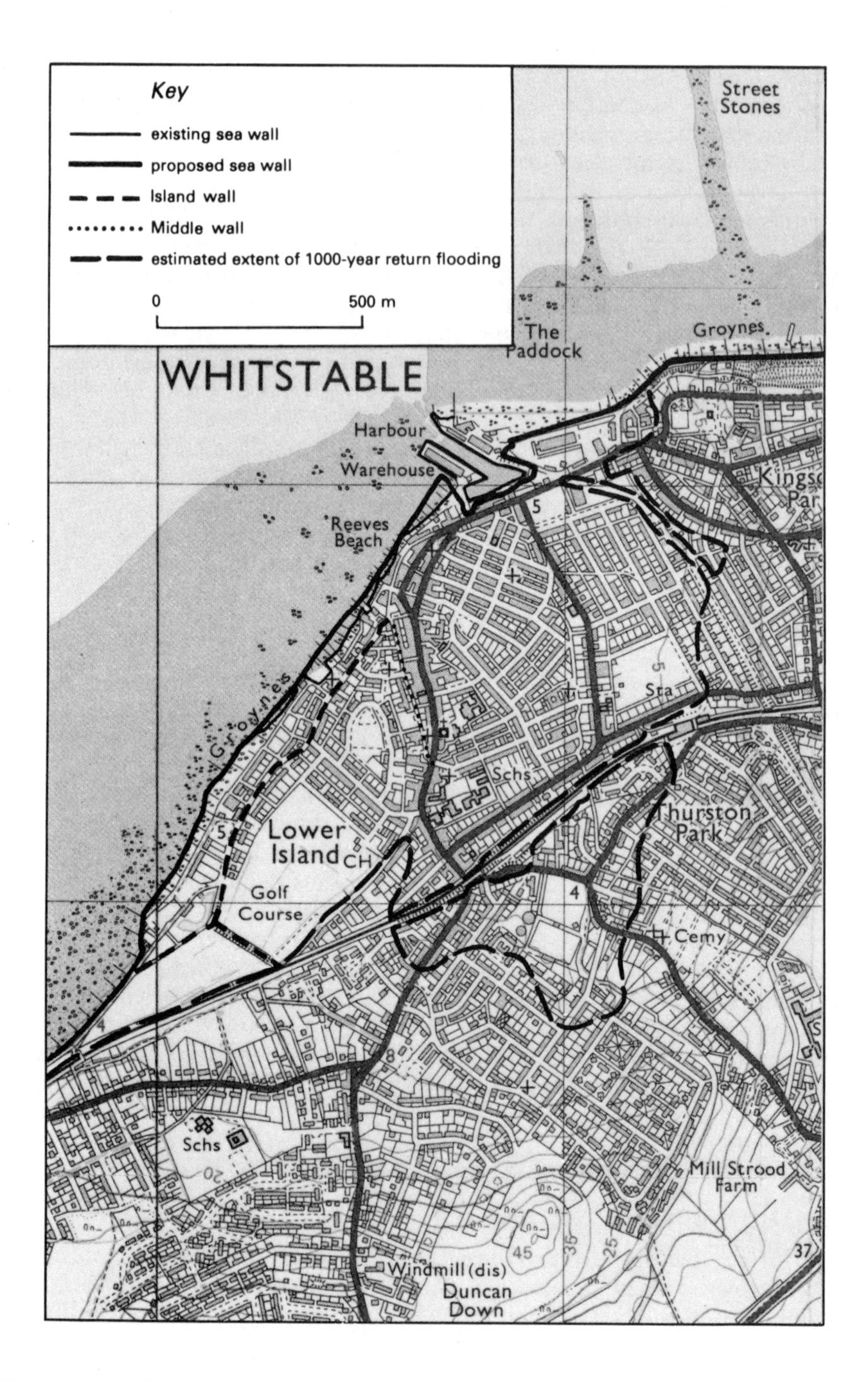

Figure 6.2 The estimated extent of sea flooding and the alignment of the proposed sea wall at Whitstable, Kent in 1981 (Crown copyright reserved).

sea levels are rising thus increasing the risk of catastrophic flooding. Built during the 1950s, the existing sea wall is now agreed by all to be structurally deficient and could be overtopped or breached resulting in heavy property damage and loss of life (Parker & Penning-Rowsell 1981b, 1982). Following a public inquiry in 1975 the City Council's proposals for a new sea wall were rejected, primarily because of their public unpopularity. Proposals for a new sea wall were subsequently devised in 1978 and full engineering details were forwarded to the Department of the Environment for approval and grant aid in March 1980.

Consultations between the City Council and central government were initially confused by a disputed division of responsibility between the Department of the Environment and the Ministry of Agriculture, Fisheries and Food. An important legal distinction exists between sea defence and coast-protection works, which affects both the level and the provenance of government grant aid. A High Court ruling was necessary to determine that the sea-wall proposals for Whitstable were primarily to prevent erosion rather than flooding. This decision gave ultimate responsibility to the Department of the Environment who, until this decision, did not require full economic appraisal of the flood protection afforded by coast-protection works.

Public consultations over the proposals appear to have occurred too little and too late. Only in August 1980 – four months after the proposals were submitted to the Department of the Environment – did the Council approve a public participation programme. This included making a planning application, mounting a public exhibition and holding a panel hearing before members and officers of the Council (Department of the Environment 1982, p. 13).

However, these consultations were complicated by ambivalent public attitudes towards flooding and by stark differences in the opinions and perceptions of City Council staff and some sections of the affected public. Support for the sea-wall proposals came from those for whom flooding would be an inconceivable personal disaster. Others also saw potential for economic development of the town once the community had invested to protect its accumulated capital in the form of domestic property and private businesses from the flood risk. As well as Council officers and some politicians these supporters included many of those most fearful of, and vulnerable to, flooding.

Opposition to the proposals came from residents and traders organised around local pressure groups such as the Whitstable Society. They believed that the Council had not learned from its earlier defeat and had still not produced a publicly acceptable proposal largely because of its failure to consult fully the local interests at a sufficiently early stage to influence the overall design. The opposition focused upon the high cost of the proposed new sea wall, the likely burden on local rates, and what they saw as an unnecessarily grandiose scheme supporting vested interests within the Council. Opposers also emphasised their lack of concern for flooding, the severity of which they disputed, as well as the disruption and risk to life from the major sea-wall construction work. The environmental impact of a substantial seafront flood defence wall was a major area of controversy and to many local interests the necessity for a public inquiry itself represented a failure in public consultation.

The tensions between opposing factions came to the fore both before and during the 1981 public inquiry. This inquiry revealed some of the common weaknesses of such public inquiries as a means of making decisions. The Council was able to employ a range of expert witnesses and studies to support its case for a sea wall, but the objectors had minimal finance and were forced to rely on the evidence of 'amateur' witnesses from a variety of professional backgrounds. Being the proposer of the scheme, and thus having a relative monopoly of the relevant data and no obligation to assess its environmental impact, the Council had a demonstrably privileged position in the advocasorial and bargaining process.

Nevertheless, after an extraordinary delay of 16 months, the Secretary of State for the Environment announced his decision to reject the Council's proposals. The reasons given were that although Whitstable's sea defences were seen as inadequate, the proposals would have severely damaged the character of the town. In addition the scheme was found to be designed to an inappropriately high standard such that costs exceeded benefits. This decision thus largely vindicated the views of local groups who had opposed the Council's scheme.

The case demonstrates the problems with flood alleviation proposals where there is inadequate public support. The Council was not legally obliged to consult local interests any more than it did, but what occurred proved inadequate. This consolidated the continued public unpopularity of the proposals. Unlike the Halvergate case, the principal interests were prepared to engage in public 'set-piece' confrontations. Private discussions and bargaining were rejected owing to mutual distrust of those concerned. A public inquiry was probably only held because it was forced on the Council through the need for compulsory purchase powers. Ultimately the Council believed some objectors to be mischievous in their intentions. Some objectors believed the Council to be arrogant in its continued pursuit of a grandiose scheme and stubborn in its belief of its correctness.

Following its latest defeat Canterbury City Council engaged a consulting engineering firm in 1983 to design a sea-defence scheme to which there would be no objectors, thus effectively itself backing away from the problem. The lesson that full and early consultation is essential appears now to have been learned and the consultants began their work with extensive local discussions. Nevertheless uncertainty remains. It may well prove impossible to avoid all objections, particularly if compulsory purchase orders are still required, if the costs are high, and if public interest in the scheme wanes while the complex design process proceeds.

Assessment

Adequate consultations are crucial to successful flood alleviation and agricultural land drainage as they are ultimately to government in any democratic society. Many interests are affected by drainage proposals and consultation needs to seek reconciliation between these views and interests, which often may be diametrically opposed. Yet such consultation – both between drainage authorities and between statutory authorities, voluntary organisations and the general public – presents considerable problems for the engineer. Rather than leading to a happy consensus, consultation and public participation can reveal previously hidden and divisive policy consequences. These can lead in the short term to polarised positions and extended political conflict which may make a consensus impossible.

For these reasons engineers and planners may be reluctant to involve the public in decisions because the public is technically uninformed or because they wish to avoid 'stirring up trouble'. Alternatively the public may be involved but often only on the engineers' terms: shades here of Arnstein's 'manipulation', 'therapy' or other degrees of tokenism (see Table 6.1). Crucial decisions may already have been taken by those with the power. Policies may be decided and only subsequently presented or defended in public. In practice, engineers and planners may well find that despite attempts to stimulate involvement the public remains uninterested in drainage proposals until the next damaging flood arrives, when confrontation reoccurs.

'Ideal' public consultation programmes therefore offer no panacea in the disputes and confrontations which sometimes arise over land drainage. Further modification of the law may be desirable over conflicts between agricultural and conservation interests, perhaps involving some element of land-use planning control so that decisions are both public and locally accountable. Legislation may also be required in the still confused area of urban drainage. Much ultimately can depend, however, upon the attitudes of the engineers or planners involved, who can either promote secrecy and the protection of the *status quo* or can demonstrate their public accountability by leading in attempts to diffuse distrust and ensure that key issues are publicly aired and that conflict is recognised.

Statutory land drainage procedures, in effect, provide only for minimum standards of consultation. In many cases this 'minimum standard' proves to be totally inadequate, and extensive voluntary consultation procedures must be developed and genuine opportunities created for the public to influence decisions. Statutory procedures are thus only the beginnings of consultation which may involve protracted private bargaining in search of a compromise. However, such a compromise may not be possible, as the Halvergate issue shows, and indeed the resolution of conflict may not be in the interests of all. Conflict resolution through negotiation and bargaining may simply result in secret 'deals' which adversely affect third parties and thus the whole political process may simply reinforce the position of those who originally held the reins of power.

7 *Fundamentals*

British flood alleviation and land drainage clearly have considerable strengths as well as a number of shortcomings. A major strength is the skill of the engineer in adapting and perfecting drainage designs based on decades or centuries of research, development and patiently accumulated experience. Furthermore, those concerned with decisions, including farmers and landowners, have a deep knowledge of drainage problems and techniques. Without this skill, knowledge and planning many towns and cities in Britain would remain as flood-prone as those elsewhere in the world which have developed uncontrolled in a free enterprise ethic or in ignorance without consideration as to flood risk. In addition, without this experience complementing moves towards regulated agricultural markets, many areas in Britain would remain as agriculturally unproductive today as they were in the economic depression of the 1920s before the report of the Royal Commission on Land Drainage in England and Wales (1927) which led to the Land Drainage Act 1930.

Our analysis of flood hazard reduction, and the inextricably linked policies for agricultural improvement through drainage, must therefore be set against comparisons with other countries and against the context of history without which any understanding will be shallow and any recommendations will be naïve.

Investigations in many disciplines over the past 25 years have enormously assisted our understanding (Parker & Penning-Rowsell 1983). Most notable in Britain has been the pioneering hydrological research undertaken at the Institute of Hydrology, the Hydraulics Research Station, the Meteorological Office and the Field Drainage Experimental Unit (Natural Environment Research Council 1975, Sutcliffe 1978, Field Drainage Experimental Unit (various dates)). The Local Government Operational Research Unit (1971, 1973, 1978) has clarified many economic issues. Research on flood damages and warnings has developed at Middlesex Polytechnic (Penning-Rowsell & Chatterton 1977, 1980, Penning-Rowsell *et al.* 1978, 1983, Green *et al.* 1983b, Parker *et al.* 1986) building on the conceptual work by the Chicago School of geographers (White & Haas 1975) and their followers (Williams 1964, Foster 1980). Flood perception studies (Penning-Rowsell 1976, Parker & Harding 1979, Smith & Tobin 1979) have usefully demonstrated many similarities with results from the USA (Burton *et al.* 1978) and Australia (Smith *et al.* 1979). Water Authorities have also contributed to British research with hydrometric studies and the consolidation of flooding and drainage records in their Section 24(5) survey reports (Parker & Penning-Rowsell 1981a). Research by government agencies, and the survey work of local and voluntary organisations, has defined incontrovertibly the potential effect of land drainage on nature conservation values (e.g. Ratcliffe 1977, Royal Society for the Protection of Birds 1983).

Quite significant problems nevertheless remain in virtually all areas of land drainage. To ignore these is both dishonest and counterproductive although to suggest, as researchers are wont to do, that further research will 'solve' all these problems is also disingenuous. More comprehensive plans for further research are available elsewhere for both the USA (Changnon *et al.* 1983) and Britain (Parker & Penning-Rowsell 1983). The discussions below are therefore restricted to certain of the more fundamental problems.

Fundamental technical problems

THE QUESTION OF 'ENCROACHMENT'

Fundamental questions remain about the extent to which floodplains should be developed and utilised. Conventional wisdom largely holds that flood protection is worthwhile when, by preventing damage, it allows intensification of land use. On the other hand, hitherto unprotected floodplains should be kept free from 'encroachment' by damageable uses. However, flood protection usually leads to further intensification of floodplain use and thus possibly raises annual average flood losses because of the effects of catastrophic floods.

Attempts are being made to control floodplain development largely because encroachment is now recognised as a fundamental cause of increasing flood loss potential in countries such as the USA, Canada and Australia. Nevertheless, the extent, pace and significance of encroachment in Britain remains under-researched. The net costs of floodplain land-use control where land for development is scarce are largely unknown. They may be far greater than seems likely since vital resource exploitation opportunities may be forgone by such restrictions.

The forces promoting encroachment are extraordinarily complex and powerful and the difficulties of controlling floodplain use have been underestimated (Burby & French 1981). In North America and Australia, for example, these forces include those promoting individual freedom, private wealth generation and local community gain. In developing countries the forces encouraging intensified floodplain land use include the need to increase food production for individuals' survival. Thus floodplain land-use controls can either require setting aside deeply-held values, by sacrificing individual freedom for some 'common good', or jeopardising the livelihood of those affected. Nevertheless attempts to increase the use of flood-prone areas should be in recognition of the full intangible effects of floods, not just the measurable damages. This, however, is highly problematic because these intangibles remain inadequately researched and difficult to assess.

The lesson, nevertheless, is that resources flow from using hazard-prone locations. Sometimes these resource gains outweigh the problems that inevitably come from hazard zone encroachment although little is known of the balance of advantage and disadvantage. This may be at least partly because, as Hewitt (1983) comments, the conventional view of a disaster or hazard is akin to ancient views of madness or other deviancy – to be separated and eradicated for fear of destabilising 'ordinary' life and 'normal' society–environment relations. The balance of advantage therefore remains uninvestigated and the conventional wisdom of avoiding development on floodplains remains pervasive.

UNCERTAINTY AND INSTABILITY

Both project appraisal results and the design of flood alleviation and land drainage schemes crucially depend upon assumptions about the continuation into the future of the events and trends that we observe today. As with all predictions, however, there is no certainty or way of knowing that stability will continue over the long life of such schemes.

If floods were not to occur with such frequency in the future as they have in the past, perhaps owing to land use or climatic change (Perry 1981), then any expenditure based on past records or present flood extents may well turn out not to have been worthwhile. If the reverse is the case, and floods occur more frequently than is foreseen, then the design of flood-control works and dam spillways could be seriously deficient and perhaps even dangerously so.

Economic instability also profoundly affects the worthwhileness of community investment in flood alleviation and agricultural drainage. Most significant here are the prices obtained for arable and livestock products which in Britain, as we have seen, are liable to political influence within the European Community's Common Agricultural Policy and therefore to unpredictable change. Currently wheat prices are high, but within the last 10 years – in Britain at least – there have been major price fluctuations reflecting changing supply and demand, the fluctuating pattern of subsidies, and the state of world markets. The world economy itself appears to become less rather than more stable, reflecting fluctuations resulting from currency speculation and such resource cost shocks as rises in oil prices.

To make the necessary predictions for flood-alleviation design and agricultural investment over the next 30 or 50 years, given this instability, is to rush in where angels fear to tread. Furthermore, to expect the world to have some 'normal' stable state, which hazards somehow interrupt, is itself naïve given the complex laws of motion and the inherent contradictions in the economies of advanced industrialised countries. Yet for practical purposes we have no other basis than to assume the economic situation and consequent investment evaluation will remain essentially the same tomorrow as it is today, false though this is bound to be!

DECISIONS

Both hazard–response theory and decision-making models (Ch. 1) aim at improving the structure of decision making and thus, for example, at ensuring that all factors or interests are considered and every alternative reviewed towards achieving some goal or meeting set objectives. A problem here, however, is knowing whether the chosen way towards decisions is in fact optimal. The unreality of normative theory based upon the pure rationality model is generally accepted, but comparison of how decisions appear to be made with one or more such decision-making models does at least prompt a host of potentially critical and insightful questions about both the planning system and the decision-making processes. For example, when does the analysis of alternatives become unnecessarily and inefficiently repetitive? How many different forms and combinations of flood hazard adjustments should be subjected to economic analysis – with all the complications that this involves – before the 'right' decision has been taken, if such a phenomenon exists?

These, moreover, are purely technical matters. More importantly, the very notion of a 'correct' decision, and decision-making system, implies the absence of distributional consequences of decisions or the absence of 'losers'. This we can now see is false, and that many decisions are fundamentally political in nature. So what is the 'optimal' method of arriving at decisions when it is demonstrable that conflict cannot be resolved and certain interests are bound to face substantial loss or uncompensated damage? Certain decisions are also irreversible in the sense that the changes thus decided cannot be undone and the *status quo* thereby restored. Once drained, wetlands can lose their wildlife value virtually for ever. Recolonisation by bird and plant species if wetland conditions could be restored some time in the future would be slow. Many European wetlands have now been drained for agricultural purposes to an extent that irreversible loss of wetland habitats has occurred, and further drainage would be critical to the remaining landscapes and wildlife.

How long in this context should bargaining proceed and inquiries take evidence? Again, such questions assume that decision making can be 'fair'. Yet we know that considerable inequities of power and influence exist within the field of land drainage in Britain, with its emphasis on farming and landowning interests. Vested interests characterise all public investment decisions and pressure from these interests inevitably distorts the allocation of resources to community welfare projects such as flood alleviation and drainage. However, it could be argued that without these interest groups, and the experience and expertise that they embrace, there might not be the initiative and political will to devise and implement such projects. Moreover, if a satisfactory state of arriving at decisions can only follow fundamental change to society, what should we do in the meantime?

Analysis of technical matters, then, tends not to be containable as such but leads directly to political questions. Some answers to these questions can emerge from a careful analysis of who gains and who loses from resource-allocation decisions. From such analysis can come moral decisions for each person concerned about whom to support and whom to oppose. Such analysis also reveals both the political character of decision making, which is obvious, and that pragmatism must be an essential element of resource allocation for public investment on flood alleviation and land drainage given a demanding public, tight deadlines and fixed budget ceilings. However, it is then difficult for the public to evaluate and to hold accountable those making decisions when there is no clear sequence for taking such decision which is accepted as 'correct' and no consensus as to what 'should' be done!

Hazards and resources: individuals and the state

Some light can also be shed on these political questions by examining the fundamental tension that exists in the relationship between the individual and the state. Moreover, the fields of flood alleviation and agricultural improvement can perhaps provide some special insights into state/individual interrelations. This is because of the complexity of the periodicity of environmental hazards and therefore the special rôle for active state intervention, even in a free-enterprise society, to protect people from their own ignorance and poor memories. Put in another way, solutions to hazard problems based on free-market decisions tend to be inapplicable.

The individual's rights and obligations are therefore controlled by regulations and constraints set and enforced by state institutions. The tensions between the individual and the state express themselves by individuals continually seeking to maximise their own advantage, perhaps by circumventing regulations or customs. In our field this might be by attempting to develop flood-prone land for individual gain, by riparian owners neglecting their obligations, or by farmers promoting agricultural intensification through drainage in disregard of collectively derived conservation recommendations.

The tensions continue because the state in Britain acts both to encourage certain aspects of individualism, and thus the entrepreneurial maximisation of private profit, and also to restrain the excesses of individualism when this is demarcated by popular pressure as being antisocial. The state here is seen by pluralists as a neutral 'referee' between individuals, and thus adjudicating what is antisocial, rather than as an autonomous force. Alternatively, Marxist explanations see the state fundamentally as a support for the ruling classes in the country and at the period in question, in our case for private enterprise capitalism. State controls and welfare expenditure are thus seen as either a means of social control to create a pliant population or as designed to further the basic interests of the ruling classes in controlling the excesses of individual entrepreneurs when these excesses could threaten the whole ruling class position.

However, the philosophy of individualism and the debate on the rôle of government or the state goes far deeper. Hobbes in the 17th century, John Stuart Mill in the 19th century and Karl Popper in the 20th century were all philosophers who stressed the need to understand society through the rôle and actions of individuals, rather than 'collectives'. Mill saw society as no more than the collection of individuals with characteristics derived simply from the properties of individuals. Durkheim, in contrast, saw society as something larger. In his view it forms a structured pattern of relations and processes which defines the rôles and duties of individuals within that society. Analysis of the rôle of the individual must therefore make reference to the social context of which he or she is a part: a person behaves differently within a capitalist society than, say, in a feudal one.

Following Durkheim and Marx we have preferred to stress the holistic or structuralist explanations of policy making (see Fig. 1.6), rather than the individualistic emphasis to explanations of events we see in studies from the USA (Parker & Penning-Rowsell 1983) and particularly in Kates's (1970) analysis of hazard response. However, this is not to deny the rôle of the individual in British flood hazard reduction. The results shown in Figure 7.1 re-analysed from Penning-Rowsell (1972, 1976), show that individuals can and do respond with damage-reducing adjustments, given a degree of experience of flooding or a lengthy residence in a flood-prone area or a

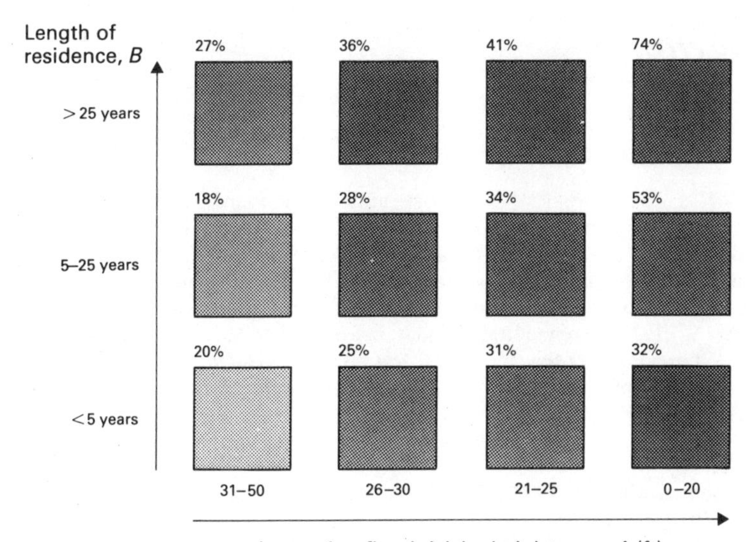

Tones show the proportion of residents adopting one damage-reducing response (the same pattern emerges for 2+ responses). The relationship indicates increasing rates of adjustment with both risk (A) and length of residence (B). When each variable is controlled by taking a single row or column, the relationships generalised by the A and B arrows still hold.

Figure 7.1 The relationship between flood hazard experience (length of residence) and adjustment in the lower Severn Valley (re-analysed from Penning-Rowsell 1972, 1976).

combination of both. Nevertheless the hazard 'problem' cannot easily be reduced for those with less experience, for whom state agencies must provide the collective memory. Furthermore, the actions of individuals are always conditioned by the institutional and social context in which they live: they may well not respond individually if state agencies will respond instead.

These examples and illustrations cannot, of course, be definitive. Indeed, we cannot settle this debate with empirical analysis or case studies because both individualistic and 'structural' explanations of policies and events can be vindicated by the same data. The dispute is about the correct starting point for constructing satisfactory explanations, and thus about methodology. A far more detailed and comprehensive analysis of policy evolution is necessary to take this debate further.

We feel, nevertheless, that more progress will be made in this further policy analysis research by starting from the holistic perspective rather than the individual's perceptions and response. This view stems from our analysis that the 'laws of motion' of flood alleviation and agricultural drainage in Britain derive largely from the overall social framework of British society rather than from individual behaviour.

Bibliography

Abrahams, M. J., J. Price, F. A. Witlock and G. Williams 1976. The Brisbane floods, January 1974; their impact on health. *Med. J. Austr.*, Dec. 18, 936–9.

Agricultural Advisory Council 1970. *Modern farming and the soil*. London: HMSO.

Agricultural Development Advisory Service 1974–7. *Getting down to drainage*. Drainage leaflets Nos. 1–21. London: Ministry of Agriculture, Fisheries and Food.

Akeroyd, A. V. 1972. Archaeological and historical evidence for subsidence in southern Britain. *Phil. Trans. R. Soc. Lond. A* **272**, 151–69.

Anglian Water Authority ND. *Milton Keynes stormwater balancing lakes*. Huntingdon: AWA.

Anglian Water Authority 1982. *Lincolnshire flood alleviation: benefit assessment*. Boston: AWA.

Anglian Water Authority 1983. *River Witham system scheme: Lincoln flood alleviation*. Boston: AWA.

Arnstein, S. 1969. A ladder of citizen participation. *J. Am. Inst. Planners* **35**, 216–24.

Bennet, G. 1970. Bristol floods 1968: controlled survey of effects on health of local community disaster. *Br. Med. J.* **3**, 454–8.

Benson, M. A. 1962. *Factors affecting the occurrence of floods in a humid region of diverse terrain*. United States Geological Survey, Water Supply Paper 1580-B. Washington: US Government Printing Office.

Bentham, J. 1980. *The Thames tidal flood problem*. Unpublished undergraduate dissertation, Department of Geography, University of Loughborough.

Beran, M. A. 1981. Recent advances in statistical flood estimation techniques. In *Flood Studies Report – five years on*. London: Institution of Civil Engineers.

Black, C. J. and J. K. Bowers 1981. *The level of protection of UK agriculture*. Discussion paper 29. Leeds: School of Economic Studies, University of Leeds.

Blenkharn, A. 1979. Rivers and land drainage. *Water Space*, Spring, 9–12.

Blenkharn, A. 1983. Conservation and land drainage: updating the guidelines. *Water bulletin* 26 August, 13–14.

Body R. 1982. *Agriculture: the triumph and the shame*. London: Temple Smith.

Bowers, J. K. 1983. Cost–benefit analysis of wetland drainage. *Environ. Plan. (A)* **15**, 227–35.

Bowers, J. K. and C. J. Black 1983. *The Soar Valley Improvement scheme: submission to the House of Lords Select Committee considering the Severn-Trent Water Authority Bill*. London: Council for the Protection of Rural England.

Boyce, D. E., N. D. Day and G. McDonald 1970. *Metropolitan plan-making*. Pennsylvania: Regional Science Research Institute, University of Pennsylvania.

Bransby-Williams, G. 1922. Flood discharge and the diversion of spillways in India. *Engineer* **134**, 321–2.

British Trust for Conservation Volunteers 1976. *Waterways and wetlands*. London: British Trust for Conservation Volunteers.

Brooks, E. 1974. Government decision-taking. *Trans. Inst. Brit. Geog.* **63**, 29–40.

Buckwell, A. E., D. R. Harvey, K. A. Panton and K. J. Thomson 1981. Some development options for the Common Agricultural Policy. *J. Agric. Econ.* **32**(3), 303–16.

Burby, R. J., and S. P. French 1981. Coping with floods: the land use management paradox. *J. Am. Plan. Assoc.* July, 289–300.

Burton, I. and R. W. Kates 1964. The perception of natural hazards in resource management. *Nat. Res. J.* **3**, 412–41.

Burton, I., R. W. Kates and G. F. White 1968. *The human ecology of extreme geophysical events*. Natural Hazard Research Working Paper No. 1. Toronto: Department of Geography, University of Toronto.

Burton, I., R. W. Kates and G. F. White 1978. *The environment as hazard*. New York: Oxford University Press.

Butters, K. and J. J. Lane 1975. Flood alleviation on some River Thames Tributaries. *J. Inst. of Water Engrs. Sci.* **29**, 67–84.

Butters, K. and W. Tuck 1973. *Evaluation of urban flood drainage*. Paper presented to the Conference of River Authority Engineers, Cranfield.

Cadbury, C. J. 1975. Populations of swans at the Ouse Washes, England. *Wildfowl* **26**, 148–59.

Caufield, C. 1981. What can we save of the Broadlands? *New Scientist* 1 January, 28–31.

Chadwick, G. 1971. *A systems view of planning*. Oxford: Pergamon Press.

Chandler, T. J. and S. Gregory 1976. *The climate of the British Isles*. London: Longman.

Changnon, S. A. *et al.* 1983. *A plan for research on floods and their mitigation in the United States*. Champaign, Ill.: Illinois State Water Survey.

Chapman, E. J. K. and R. W. Buchanan 1966. Frequency of floods of normal maximum intensity in upland areas of Great Britain. In *River flood hydrology*, 65–86. London: Institution of Civil Engineers.

Chatterton, J. B. 1983. *Proof of evidence*. House of Lords Select Committee on the Severn Trent Water Authority Bill. Birmingham: Severn Trent Water Authority.

Chatterton, J. B. and S. J. Farrell 1977. *Nottingham flood warning scheme: Benefit assessment*. London: Middlesex Polytechnic Flood Hazard Research Project.

Chatterton, J. B. and T. Lau 1983. *Discounted cash flow (DCF) model for appraisal of land drainage and flood alleviation schemes: a user manual*. Birmingham: Severn Trent Water Authority.

Chatterton, J. B. and E. C. Penning-Rowsell 1978. The benefits of urban storm drainage: computer modelling and standardised assessment techniques. In *Proceedings of the International Conference on Storm Drainage*, Southampton 1978, 648–65.

Chatterton, J. B. and E. C. Penning-Rowsell 1981. Computer modelling of flood alleviation benefits. *J. Water Resources Plan. Man. Div., Am. Soc. Civil Engrs* **107** (WR2), 533–47.

Checkley, K. 1982, *Finance for farming*. London: Institute of Bankers.

Chow, V. T. 1959. *Open channel hydraulics*. New York: McGraw-Hill.

Chow, V. T. (ed.) 1964. *Handbook of applied hydrology*. New York: McGraw-Hill.

Clark, M. A. 1982. *Towards a landscape strategy for the Broads*. Report BA SMP6. Norwich: Broads Authority.

Cole, G. 1976. Land drainage in England and Wales. *J. Inst. Water Engrs Sci.* **30**(7), 345–67.

Cole, G. and E. C. Penning-Rowsell 1981. The place of economic evaluation in determining the scale of flood alleviation works. In *Flood Studies Report – 5 years on*, 143–51. London: Institution of Civil Engineers.

Cordery, J. 1971. Estimation of design hydrographs for small rural catchments. *J. Hydrol.* **13**, 263–77.

Cotgrove, S. 1982. *Catastrophe or cornucopia: the environment, politics and the future*. Chichester: Wiley.

Council for the Protection of Rural England 1983. *Severn-Trent Water Authority Bill: Memorandum to the House of Lords Select Committee*. London: CPRE.

Craeger, W. P. 1945. *Engineering for dams*, vol. 1. New York: Wiley.
Cullingworth, J. B. 1972. *Town and country planning in Britain*, 4th ed. London: George Allen & Unwin.
Cunge, J. A. 1969. On the subject of a flood propagation method. *J. Hydrol. Res.* **7**, 205–30.

Dalrymple, T. 1960. Flood frequency analysis. In *Manual of Hydrology*, Part 3, Flood-flow techniques. United States Geological Survey, Water Supply Paper 1543-A. Washington, DC: US Government Printing Office.
Darby, H. C. 1940. *The draining of the Fens*. Cambridge: Cambridge University Press.
Darby, H. C. 1983. *The changing fenland*. Cambridge: Cambridge University Press.
Davies, I. and D. J. Parker 1982. *Water quality control and urban angling interests*. Paper presented at the Institute of British Geographers Annual Conference, Southampton.
Departmental Committee on Coastal Flooding 1954. *Cmnd 9165*. London: HMSO.
Department of National Development and Energy (Australia) 1981. *Proceedings of the floodplain management conference*. Australian Water Resources Council Conference Series No. 4. Canberra: Australian Government Publishing Service.
Department of the Environment 1982. *Canterbury City Council, applications and appeals, sea defences, Whitstable, County of Kent*. London: DoE.
Department of the Environment, Ministry of Agriculture, Fisheries and Food, Welsh Office Agriculture Department 1982. *Land drainage and conservation: guidance notes on procedures for Water Authorities, Internal Drainage Boards, Nature Conservancy Council and Countryside Commission. Section 22 of the Water Act 1973 as amended by Section 48 of the Wildlife and Countryside Act 1981*. London: DoE *et al*.
Department of the Environment, Welsh Office, Ministry of Agriculture, Fisheries and Food 1977. *The water industry in England and Wales: the next steps*. London: HMSO.
Dror, Y. 1964. Muddling through – science or inertia? *Public Admin. Rev.* **24**(3).
Duffey, E. 1974. *Nature reserves and wildlife*. London: Heinemann.
Dunham, K. C. and D. A. Gray (eds) 1972. A discussion on problems associated with the subsidence of South Eastern England. *Phil. Trans. R. Soc. Lond. A* **272**, 79–274.

Eckstein, O. 1968. *Water resource development: the economics of project evaluation*. Cambridge, Mass.: Harvard University Press.
Edwards, M. 1977. *The ideological function of cost–benefit analysis in planning*. Discussion paper 25. London: School of Environmental Studies, University College London.
Etzioni, A. 1967. Mixed-scannings: a 'third' approach to decision-making. *Public Admin. Rev.* **27**, 385–92.
European Commission 1981. *The Common Agricultural Policy*. Brussels: Commission of the European Communities.

Faludi, A. 1973. *A reader in planning theory*. Oxford: Pergamon Press.
Farquharson, F. A. K., M. J. Lowing and J. V. Sutcliffe 1975. Some aspects of design flood estimation. In *Symposium proceedings, BNCOLD Symposium*. Newcastle: Newcastle University/BNCOLD.
Field Drainage Experimental Unit (various dates). *Technical Bulletins*. London: Ministry of Agriculture, Fisheries and Food.
Flather, R. A. 1980. *Practical surge prediction using numerical models*. Paper read at

Bristol conference on Floods Due to High Tides and Winds, Institute of Mathematics and its Applications.

Foster, H. D. 1980. *Disaster planning: the preservation of life and property*. Berlin: Springer Verlag.

Francis, J. R. D. 1973. Rain, run-off and rivers. *Q. J. R. Meteorol. Soc.* **99**, 556–68.

Freeman, L. 1981. Fighting the surge. *Water* **36**, 2–7.

Garner, J. F. 1980. *The law of sewers and drains*, 6th edn. Croydon: D. R. Publications.

George, M. 1977. The decline in Broadland's aquatic fauna and flora. *Trans. Norfolk Norwich Naturalists' Soc.* **24**, 42–53.

Gilman, K. n.d. *Nature conservation in wetlands – two small fen basins in Western Britain*. Wallingford, Oxford: Institute of Hydrology.

Gilman, K. and M. D. Newson n.d. *The Anglesey wetlands study*. Wallingford, Oxford: Institute of Hydrology.

Greater London Council 1970. *Thames flood prevention: first report of studies, Department of Public Health Engineering*. London: GLC.

Greater London Council 1971. *Thames flood prevention: second report of studies, Department of Public Health Engineering*. London: GLC.

Green, C. H., D. J. Parker and D. J. Emery 1983a. *The real costs of flooding to households: the intangible costs*. London: Middlesex Polytechnic.

Green, C. H., D. J. Parker, P. Thompson and E. C. Penning-Rowsell 1983b. *Indirect losses from urban flooding: an analytical framework*. Geography and Planning Working Paper 6. London: Middlesex Polytechnic.

Green, C. H. and E. C. Penning-Rowsell 1983. *The estimation of the opportunity costs of labour and agricultural products, and the choice of the appropriate Test Discount Rate*. London: Middlesex Polytechnic Flood Hazard Research Centre.

Green, F. H. W. 1979. Field under-drainage and the hydrological cycle. In *Man's impact on the hydrological cycle in the UK*, G. E. Hollis (ed.), 9–17. Norwich: Geobooks.

Green, F. H. W. 1980. Current field drainage in northern and western Europe. *J. Environ. Man.* **10**, 149–53.

Gumbel, E. J. 1941. The return period of flood flows. *Ann. Math. Stat.* **12**, 143–90.

Gumbel, E. J. 1958. *Statistics of extremes*. New York: Columbia University Press.

Gutch, R. 1972. *Goals and the planning process*. Oxford working papers in planning education and research No. 11. Oxford: Department of Town Planning, Oxford Polytechnic.

Hall, C. 1978. Amberley Wild Brooks. *Vole* **7**, 14–15.

Hall, D. G. M., M. J. Reeve, A. J. Thomasson and V. F. Wright 1977. *Water retention, porosity and density of field soils*. Soil Survey Technical Monograph No. 9. Harpenden, Hertfordshire: The Soil Survey of England and Wales.

Hall, M. J. 1980. A historical perspective on the Flood Studies Report. In *Flood Studies Report – Five years on*, 11–17. London: Institution of Civil Engineers.

Hallas, P. S. 1981. Experience in the use of the Flood Studies Report for reservoir spillway design. In *Flood Studies Report – Five years on*, 79–83. London: Institution of Civil Engineers.

Hamnett, S. L. 1973. *Goals as aids to justification: some implications for rational planning*. Oxford working papers in planning education and research No. 17. Oxford: Department of Town Planning, Oxford Polytechnic.

Handmer, J. W. and D. I. Smith 1983. Health hazards of floods: hospital admissions for Lismore. *Austr. Geogr. Stud.* **21**, 221–30.

Harding, D. M. and D. J. Parker 1974. Flood hazard at Shrewsbury, United Kingdom. In *Natural hazards local, national, global*, G. F. White (ed.), 43–52. New York: Oxford University Press.

Harvey, D. 1969. *Explanation in geography*. London: Edward Arnold.
Haslam, S. M. 1973. The management of British wetlands. I. Economic and amenity use. *J. Environ. Man.* **1**, 303–20.
Hayami, S. 1951. *On the propagation of flood waves*. Bulletin No. 1. Disaster Prevention Research Institute. Kyoto University, Japan.
Hayashi, T. 1969. *Propagation and deformation of flood waves in natural channels*. Anniversary Bulletin, 67–80, Chuo University, Japan.
Henderson, F. M. 1966. *Open channel flow*. New York: Macmillan.
Heras, R. 1969. Méthodes practiques d'estimation des plus grandes crues. In *Floods and their computation*, Proceedings of the Leningrad Symposium 1967. IASH/UNESCO/WMO Publication, 492–504. Paris: UNESCO.
H.M. Treasury 1982. *Investment appraisal in the public sector*. London: H.M. Treasury.
Hewitt, K. (ed.) 1983. *Interpretations of calamity*. London: George Allen & Unwin.
Hill, A. R. 1976. The environmental impact of agricultural land drainage. *J. Environ. Man.* **4**, 251–74.
Hinge, D. C. and G. E. Hollis (eds.) 1980. *Land drainage, rivers, riparian areas and nature conservation*. Department of Geography Conservation Discussion Paper. London: University College London.
Hitchenor, J. P. 1980. The local authorities and land drainage: a county view point. Paper presented at Seminar on *Land drainage – whose responsibility?* 23 October. London: Institution of Municipal Engineers.
Hodgson, J. M. 1974. *The soil survey field handbook*. Harpenden, Hertfordshire: The Soil Survey of England and Wales.
Hollis, G. E. 1975. The effect of urbanisation on floods of different recurrence intervals. *Water Resources Res.* **11**, 431–34.
Hollis, G. E. (ed.) 1979. *Man's impact on the hydrological cycle in the U.K.* Norwich: Geobooks.
Hollis, G. E. 1980. Land drainage and nature conservation: is there a way ahead? *Ecos* **3**(1), 3–11.
Horner, R. W. 1978. Thames tidal flood works in the London excluded area. *J. Inst. Publ. Health Engrs* **6**(1), 16–24.
Horner, R. W. 1979. The Thames barrier project. *Geog. J.* **145**(2), 242–53.
House of Lords Select Committee on the Severn Trent Water Authority Bill 1983. *Special report*. London: HMSO.
Howe, C. W. 1971. *Benefit–cost analysis for water system planning*. Water resources monography 2. Washington: American Geophysical Union.
Howe, C. W. 1977. *The design and evaluation of institutional arrangements for water planning and management*. Paper presented at the UN Water Conference, Mar Del Plata 14–25 March.
Howe, G. M., H. O. Slaymaker and D. M. Harding 1967. Some aspects of the flood hydrology of the upper catchments of the Severn and Wye. *Trans. Inst. Brit. Geogs* **41**, 33–58.
Hydraulics Research Station n.d. *River flow prediction using computer models*. Wallingford, Oxford: HRS.

IASH/UNESCO/WMO 1969. *Floods and their computation*. Proceedings of the Leningrad Symposium, 1967. Paris: UNESCO.
Institute of Hydrology 1977a. *Areal reduction factor in flood frequency estimation*. Flood studies supplementary report No. 1. Wallingford, Oxford: IOH.
Institute of Hydrology 1977b. *The estimation of low return period floods*. Flood studies supplementary report No. 2. Wallingford, Oxford: IOH.
Institute of Hydrology 1977c. *A report on the seminar 'The flood studies report – an*

opportunity for discussion' held at Birmingham University March 1977. Flood studies supplementary report No. 3. Wallingford, Oxford: IOH.

Institute of Hydrology 1977d. *Some results of a search for historical information on chalk catchments.* Flood studies supplementary report No. 4. Wallingford, Oxford: IOH.

Institute of Hydrology 1978a. *Flood prediction for small catchments.* Flood studies supplementary report No. 6. Wallingford, Oxford: IOH.

Institute of Hydrology 1978b. *Revised winter rain acceptance potential.* Flood studies supplementary report No. 7. Wallingford, Oxford: IOH.

Institute of Hydrology 1978c. *A comparison between the national formulae and the flood studies unit hydrograph procedure.* Flood studies supplementary report No. 8. Wallingford, Oxford: IOH.

Institute of Hydrology 1979. *Design flood estimation in catchments subject to urbanisation.* Flood studies supplementary report No. 5. Wallingford, Oxford: IOH.

Institution of Civil Engineers 1933. *Committee on floods in relation to reservoir practice. Interim report.* London: ICE.

Institution of Civil Engineers 1960. *Floods in relation to reservoir practice.* London: ICE.

Institution of Civil Engineers 1967. *Flood studies for the United Kingdom.* London: ICE.

Institution of Civil Engineers 1969. *An introduction to engineering economics.* London: ICE.

Institution of Civil Engineers 1981. *Flood studies report – five years on.* Proceedings of a conference held in Manchester, July 1980. London: ICE.

James, L. D. and R. R. Lee 1971. *Economics of water resource planning.* New York: McGraw-Hill.

Janis, I. L. and L. Mann 1977. Emergency decision making: a theoretical analysis of response to disaster warnings. *J. Human Stress* June, 35–48.

Johnson, E. A. G. 1954. Land drainage in England and Wales. *Proc. Inst. Civil Engrs* Part III, **3**(3), 601–51.

Johnson, F. G., R. M. Jarvis and G. Reynolds 1981. Use made of the flood studies report for reservoir operation in hydroelectric schemes. In *Flood Studies Report – five years on*, 85–90. London: ICE.

Kasperson, R. E. 1969. Environmental stress and the municipal political system. In *The structure of political geography*, R. E. Kasperson and J. V. Minghi (eds.), 481–96. Chicago: Aldine.

Kates, R. W. 1962. *Hazard and choice perception in floodplain management.* Research paper no. 78. Chicago: Department of Geography, University of Chicago.

Kates, R. W. 1965. *Industrial flood losses: damage estimation in the Lehigh Valley.* Research paper no. 98. Chicago: Department of Geography, University of Chicago.

Kates, R. W. 1970. *Natural hazard in human ecological perspective: hypotheses and models.* Natural Hazard Research Working Paper No. 14. Toronto: Department of Geography, University of Toronto.

Kavanagh, N. J. and J. R. Slater 1975. *Final report on river Frome, Gloucestershire, improvement scheme: benefit evaluation.* Birmingham: University of Birmingham.

Kelcey, J. 1982. *Urban Ecology No. 6. Ecological and recreational aspects of the design of wet balancing lakes.* Milton Keynes: Milton Keynes Development Corporation.

Land Drainage Act 1976. London: HMSO.

Land Use Consultants 1978. *A landscape assessment of the Yare Basin flood control study proposals*. Countryside Commission working paper 13. Cheltenham: Countryside Commission.

Law, F. M. 1981. The flood studies report and the Institution's guide to floods and reservoir safety. In *Flood Studies Report – five years on*, 71–77. London: ICE.

Lawler, E. A. 1964. Flood routing. In *Handbook of applied hydrology*, V. T. Chow (ed.) New York: McGraw-Hill.

Learmouth, A. 1950. *The floods of 12th August 1948 in South-east Scotland*. Unpublished paper. Liverpool: University of Liverpool.

Lewin, J. 1981. *British rivers*. London: George Allen & Unwin.

Lewin, J. and D. Hughes 1980. Welsh floodplain studies – II. Application of a qualitative inundation model. *J. Hydrol.* **46**, 35–49.

Lewis, J. 1979. *Vulnerability to a natural hazard: geomorphic, technological and social change at Chiswell, Dorset*. Natural hazard research working paper no. 37. Boulder: University of Colorado.

Lindblom, E. C. 1959. The science of 'muddling through'. *Publ. Admin. Rev.* **19**, 79–88.

Lloyd-Davis, E. E. 1906. The elimination of storm water from sewerage systems. *Proc. Inst. Civil Engrs* **164**, 41–67.

Local Government Operational Research Unit 1971. *Cost–benefit analysis of Towcester flood relief scheme*. Report T33. Reading: LGORU.

Local Government Operational Research Unit 1973. *The economics of flood alleviation*. Report C155. Reading: LGORU.

Local Government Operational Research Unit 1978. *The economics of pumped drainage*. Report C271. Reading: LGORU.

Lowing, M. H. and D. W. Reed 1981. Recent advances in flood estimation techniques based on rainfall-runoff. In *Flood studies report – five years on*, 49–56. London: Institution of Civil Engineers.

McCarthy, G. T. 1938. *The unit hydrograph and flood routing*. Unpublished paper presented at a conference of the North Atlantic Division of the U.S. Army Corps of Engineers.

MacDonald and Partners 1978. *The River Hull tidal surge barrier*. Paper delivered to the conference of River Engineers, Cranfield, July 1978.

McDonald, A. and D. Ledger 1981. *Flood area modelling from an elementary data base*. Leeds: School of Geography, University of Leeds.

Mackey, P. G. 1981. Use made of the flood studies report for reservoir operation for water supply and flood control. In *Flood Studies Report – five years on*, 91–106. London: ICE.

McLoughlin, J. B. 1969. *Urban and regional planning: a systems view*. London: Faber.

Manley, G. 1952. *Climate and the British scene*. London: Collins.

Mantz, P. A. and H. L. Wakeling 1979. Forecasting flood levels for joint events of rainfall and tidal surge flooding using extreme value statistics. *Proc. Inst. Civil Engrs* **67**(2), 31–50.

Marshall, J. K. 1977. Use of the Flood Studies Report for a drainage study at Hereford. *J. Inst. Water Engrs Scientists* **31**(3), 187–201.

Meteorological Office 1973. *British rainfall 1967*. London: HMSO.

Miers, R. H. 1967. A wide view of land drainage. *Association of River Authorities Yearbook*, 56–71.

Miers, R. H. 1979. Land drainage – its problems and solutions. *J. Inst. Water Engrs Scientists* **33**, 547–79.

Mileti, D. S. 1975. *Natural hazard warning systems in the United States: a research assessment*. Program on technology, environment and man. Boulder, Colorado: Institute of Behavioural Science, University of Colorado.

Ministère de d'Equipment et du Logement 1969. *Recherches méthodologiques sur la rentabilité économique des mesures de controle des crues a l'etranger*. Paris: Direction des Ports Maritimes et des Voies Navigables, Service Centrale Hydrologique.

Ministry of Agriculture, Fisheries and Food 1974. *Guidance notes for Water Authorities, Water Act 1973, section 24*. London: MAFF.

Ministry of Agriculture, Fisheries and Food 1978a. *Cost–benefit analysis for flood protection and land drainage projects: supplementary note on agricultural benefit*. London: MAFF.

Ministry of Agriculture, Fisheries and Food 1978b. *The review of the land drainage powers of Water Authorities and local authorities: consultation paper*. London: MAFF.

Ministry of Agriculture, Fisheries and Food 1979. *Farming and the nation*. Cmnd. 7458. London: HMSO.

Ministry of Agriculture, Fisheries and Food 1980. *Agriculture and horticulture development scheme. Explanatory leaflet for grants under approved development plans*. Leaflet AHS 5. London: MAFF.

Ministry of Agriculture, Fisheries and Food 1981a. *Agriculture and horticulture grant scheme (AHGS) agriculture and horticulture development scheme (AHDS)* Leaflet AHS 23. London: MAFF.

Ministry of Agriculture, Fisheries and Food 1981b. *Conference of river engineers – report 1981* 7–9 July. London: Land and Water Service, MAFF.

Ministry of Agriculture, Fisheries and Food 1982. *Agriculture and horticulture grant scheme explanatory leaflet for investment grants*. Leaflet AHS 2. London: MAFF.

Mitchell, B. 1971. *Water in England and Wales: supply, transfer and management*. Research Paper No. 9. Liverpool: Department of Geography, University of Liverpool.

Mishan, E. J. 1971. *Cost–benefit analysis*. London: George Allen & Unwin.

Moore, N. 1980. When there are no butterflies left. *Guardian* 10 January, 11.

Morris, J. and Hess, T. M. 1984. *Drainage benefits and farmer uptake*. Paper to Cranfield conference of river engineers, Silsoe College, Beds.

Nash, J. E. 1960. A unit hydrograph study, with particular reference to British catchments. *Proc. Inst. Civil Engrs* **17**, 249–82.

National Farmers' Union and Country Landowners' Association 1977. *Caring for the countryside: a statement of intent for farmers and landowners*. London: NFU and CLA.

National Water Council 1978. *Water industry review 1978*. London: NWC.

National Water Council 1981. *Design and analysis of urban storm drainage: the Wallingford Procedure*, 4 vols. Department of the Environment/National Water Council Standing Technical Committee Report 28. London: NWC.

National Water Council 1982. *Water industry review 1982, supporting analysis*. London: NWC.

Natural Environment Research Council 1975. *Flood studies report*, vols. I–V. London: NERC.

Nature Conservancy Council 1977. *Nature conservation and agriculture*. London: NCC.

Nature Conservancy Council, South West Region 1977. *The Somerset Levels wetlands project: a consultation paper*. Taunton: NCCSWR.

Newbold, C. 1982. The management principles of nature conservation and land drainage – two antagonistic aims? In *Proceedings EWRS 6th Symposium on Aquatic Weeds*. London: NCC.

Newbold, C., J. Pursglove and N. Holmes 1983. *Nature conservation and river engineering*. London: NCC.

Newson, M. D. 1975. *Flooding and flood hazard in the United Kingdom*. Oxford: Oxford University Press.

Newson, M. D. and J. G. Harrison 1978. *Channel studies in the Plynlimon experimental catchments*. Report No. 47. Wallingford, Oxford: IOH.

Northumbrian Water Authority 1978. *Water Act 1973, section 24/5 land drainage survey*, 4 vols. Gosforth: NWA.

Nicholson, M. 1972. *The environmental revolution: a guide to the new masters of the Earth*. London: Penguin.

Nix, J. 1980. *Farm management pocketbook*, 11th ed. Wye, Kent: Wye College, University of London.

Oates, F. L. (South West Water Authority) 1981. Personal communication quoted in Van Oosterom (1981).

Odum, E. P. 1971. *Fundamentals of ecology*. 3rd ed. Philadelphia: W. B. Saunders.

Office of Public Works (Eire) 1974. *River Maigue drainage scheme: cost–benefit analysis*. Prl. 5842. Dublin: OPW.

O'Riordan, T. 1971. *Perspectives on resource management*. London: Pion.

O'Riordan, T. 1976. *Environmentalism*. London: Pion.

O'Riordan, T. 1979. The Broads Authority. *Town Country Plan*. May, 49–52.

O'Riordan, T. 1980a. *Lessons from the Yare Barrier Controversy*. Norwich: School of Environmental Sciences, University of East Anglia.

O'Riordan, T. 1980b. A case study in the politics of land drainage. *Disasters* **4**(4), 393–410.

O'Riordan, T. 1980c. Book review of 'The benefits of flood alleviation: A manual of assessment techniques.' *Disasters* **4**(2), 251.

O'Riordan, T. 1980d. The Yare Barrier proposals. *Ecos* **1**(2), 8–14.

O'Riordan, T. 1981. *Environmentalism*, 2nd ed. London: Pion. See final chapter.

O'Riordan, T. and W. R. D. Sewell (eds.) 1982. *Project assessment and policy review*. Chichester: John Wiley.

Park, C. C. 1977. Man-induced changes in stream channel capacity. In *River channel changes*, K. J. Gregory (ed.), 121–44. Chichester: John Wiley.

Parker, D. J. 1976. *Socio-economic aspects of floodplain occupance*. Unpublished PhD thesis, University College of Wales.

Parker, D. J. 1981. Flood mitigation through non-structural measures: a critical appraisal. In *Proceedings of the international conference on flood disasters*, vol. 1. New Delhi: Indian Academy of Sciences.

Parker, D. J. 1983. *Some current developments and progress in flood hazard research*. Paper presented at Annual Meeting of the British Association for the Advancement of Science, Section E7, Brighton. London: Middlesex Polytechnic Flood Hazard Research Centre.

Parker, D. J., C. H. Green and P. M. Thompson 1983a. Red herrings, teapots and icebergs: are we missing some of the crucial flood alleviation benefits? In *Cranfield conference papers*. London: Middlesex Polytechnic Flood Hazard Research Centre.

Parker, D. J., C. H. Green and P. M. Thompson 1986. *Urban flood protection benefits: a project appraisal guide*. Farnborough: The Technical Press.

Parker, D. J., C. H. Green and E. C. Penning-Rowsell 1983b. *Swalecliff coast protection scheme: evaluation of potential benefits*. London: Middlesex Polytechnic Flood Hazard Research Centre.

Parker, D. J. and D. M. Harding 1979. Natural hazard perception, evaluation and adjustment. *Geography* **64**, 307–16.

Parker, D. J. and E. C. Penning-Rowsell 1980. *Water planning in Britain*. London: George Allen & Unwin.
Parker, D. J. and E. C. Penning-Rowsell 1981a. Specialist hazard mapping: The Water Authorities land drainage surveys. *Area* **13**(2), 97–103.
Parker, D. J. and E. C. Penning-Rowsell 1981b. *Whitstable Central area coast protection scheme: benefit assessment*. London: Middlesex Polytechnic Flood Hazard Research Centre.
Parker, D. J. and E. C. Penning-Rowsell 1982. Flood risk in the urban environment. In *Geography and the urban environment*, D. T. Herbert and R. J. Johnson (eds.), 201–39. Chichester: Wiley.
Parker, D. J. and E. C. Penning-Rowsell 1983. Flood hazard research in Britain. *Progr. Human Geog.* **7**(2), 182–202.
Paynting, T. 1982. Flood scheme reconciles conservation and alleviation. *Surveyor* November 4, 14–16.
Pearce, D. W. 1976. *Environmental economics*. London: Longman.
Pearce, D. W. 1977. Accounting for the future. *Futures* **9**, 365–74.
Pearce, D. W. (ed.) 1978. *Valuation of social cost*. London: George Allen & Unwin.
Pearce, D. W. and C. A. Nash 1981. *The social appraisal of projects: a text in cost–benefit appraisal*. London: Macmillan.
Penning-Rowsell, E. C. 1972. *Flood hazard research project*. Progress report 1. London: Middlesex Polytechnic.
Penning-Rowsell, E. C. 1974. Landscape evaluation for development plans. *J. R. Town Plan. Inst.* **60**(10), 930–4.
Penning-Rowsell, E. C. 1976. The effect of flood damage on land use planning. *Geogr. Polon.* **34**, 139–53.
Penning-Rowsell, E. C. 1978. *Proposed drainage scheme for Amberley Wild Brooks, Sussex: benefit assessment*. London: Middlesex Polytechnic Flood Hazard Research Centre.
Penning-Rowsell, E. C. 1980. Land drainage policy and practice: who speaks for the environment? *Ecos* **1**(3), 16–21.
Penning-Rowsell, E. C. 1981a. Non-structural approaches to flood control: floodplain land-use regulation and flood warning schemes in England and Wales. In *Proceedings of the International Commission on Irrigation and Drainage*, 11th Congress, 193–211. Grenoble.
Penning-Rowsell, E. C. 1981b. The implications of our radically over-designed urban storm sewer and drainage systems. In *Proceedings of the International Conference: Water Industry '81*, Brighton, 149–54. Edinburgh: CEP Consultants Ltd.
Penning-Rowsell, E. C. 1981c. Fluctuating fortunes in gauging landscape quality. *Progr. Human Geog.* **5**(1), 25–41.
Penning-Rowsell, E. C. 1982a. Britain under water. *Geogr. mag.* **54**(4), 184–6.
Penning-Rowsell, E. C. 1982b. Planning and water services: keeping in step. *Town Country Plan.* **51**(6), 150–2.
Penning-Rowsell, E. C. 1983a. An evaluation of wetland policy in England and Wales. In *The future of the wetlands: assessing visual-cultural values*, R. C. Smardon (ed.), 25–39. New Jersey: Allanheld, Osmon.
Penning-Rowsell, E. C. 1983b. Benefit–cost analysis: friend or foe? In *Cranfield conference papers*. London: Middlesex Polytechnic Flood Hazard Research Centre.
Penning-Rowsell, E. C. and J. B. Chatterton 1976. Constraints on environmental planning: the example of flood alleviation. *Area* **8**(2), 133–8.
Penning-Rowsell, E. C. and J. B. Chatterton 1977. *The benefits of flood alleviation: A manual of assessment techniques*. Farnborough, England: Saxon House/Gower Press.
Penning-Rowsell, E. C. and J. B. Chatterton 1980. Assessing the benefits of flood alleviation and land drainage. *Proc. Inst. Civil Engrs* **69**(2), 295–315, 1051–4.

Penning-Rowsell, E. C. and J. B. Chatterton 1984. Gauging the economic viability of agricultural land drainage schemes. *J. Inst. Water Engrs Scient.* (in press).

Penning-Rowsell, E. C., J. B. Chatterton and D. J. Parker 1978. *The effect of flood warning on flood damage reduction: a report for the Central Water Planning Unit*. Reading: CWPU.

Penning-Rowsell, E. C and D. J. Parker 1974. Improving floodplain development control. *J. R. Town Plan. Inst.* **60**, 540–3.

Penning-Rowsell, E. C. and D. J. Parker 1980. *Chesil sea defence scheme: benefit assessment*. London: Middlesex Polytechnic Flood Hazard Research Centre.

Penning-Rowsell, E. C. and D. J. Parker 1983. The changing economic and political character of water planning in Britain. In *Progress in Resource Management and Environmental Planning*, T. O'Riordan and R. K. Turner (eds.), vol. 4, 169–99. Chichester: Wiley.

Penning-Rowsell, E. C. and D. J. Parker 1984. *Water Services*. Social Science Research Council/Royal Statistical Society. Oxford: Pergamon Press.

Penning-Rowsell, E. C., D. J. Parker, D. Crease and C. R. Mattison 1983. *Flood warning dissemination: an evaluation of some current practices in the Severn Trent Water Authority area*. Geography and Planning Paper 7. London: Middlesex Polytechnic.

Perry, A. H. 1981. *Environmental hazards in the British Isles*. London: George Allen & Unwin.

Platt, R. H. 1980. *Intergovernmental management of floodplains*. Program on Technology, Environment and Man No. 30. Colorado: Institute of Behavioural Science, University of Colorado.

Platt, R. H. and G. M. McMullen 1979. *Fragmentation of public authority over floodplains: the Charles River response*. Publication 101, Water Resources Research Center. Massachusetts: University of Massachusetts.

Pollard, M. 1978. *North Sea surge: the story of the East Coast floods of 1953*. Lavenham, Suffolk: Terence Dalton.

Poots, A. D. and S. R. Cochrane 1979. Design flood estimation for bridges, culverts and channel improvements on small rural catchments. *Proc. Inst. Civil Engrs* **66**(1), 663–6.

Porter, E. A. 1970. *The assessment of flood risk for land-use planning and property management*. Unpublished PhD thesis, University of Cambridge.

Potter, W. D. 1961. *Peak rates of runoff from small watersheds*. United States Department of Commerce Bureau of Public Roads, Hydraulic Design Series No. 2. Washington: US Government Printing Office.

Potter, H. R. 1978. *The use of historic records for the augmentation of hydrological data*. Report No. 46. Wallingford, Oxford: IOH.

Powell, R. W. 1943. A simple method of estimating flood frequencies. *Civil. Engr.* **13**, 105–6.

Pratt, S. and J. M. Holloway 1978. *Thames barrier project: Provision of services*. Paper presented to Building Construction Forum, February 1978.

Price, R. K. 1973. *Flood routing methods for British rivers*. Hydraulics Research Station, Report INT-III. Wallingford, Oxford: HRS.

Price, R. K. 1975. (Revised September 1977). *A mathematical model for river flows: I. Theoretical development*. Hydraulics Research Station, Report INT-127. Wallingford, Oxford: HRS.

Price, R. K. 1977a. *FLOUT – a river catchment flood model*. Hydraulics Research Station Report No. IT 168. Wallingford, Oxford: HRS.

Price, R. K. 1977b. *Estimation of flood levels on floodplains for prescribed discharges*. Paper presented to the Conference of River Engineers, Cranfield.

Price, R. K. 1980. *FLOUT: a river catchment model*. Report IT 168, revised edition: Hydraulics Research Station. Wallingford, Oxford: HRS.

Price, R. K. and Samuels, P. G. 1980. A computational hydraulic model for rivers. *Proc. Inst. Civ. Engrs P2*, **69**, 87–96.

Quarentelli, E. L. 1980. *Evacuation behaviour and problems: findings and implications from the research literature*. Disaster Research Center Miscellaneous Report 27. Ohio: Ohio State University.

Ratcliffe, D. A. (ed.) 1977. *A nature conservation review*, 2 vols. Cambridge: Cambridge University Press.

Ratcliffe, J. B. and R. P. Hattey 1982. *Welsh lowland peatland survey*. London: NCC.

Regan, M. M. and E. G. Weitzell 1947. Economic evaluation of soil and water conservation measures and programs. *J. Farm Econ.* November, 1275–94.

Richardson, J. J., A. G. Jordan and R. H. Kimber 1978. Lobbying, administrative reform and policy styles: the case of land drainage. *Polit. Studies* **26**(1), 47–64.

Robinson, M. 1980. *The effect of pre-afforestation drainage on the streamflow and water quality of a small upland catchment*. Report No. 73. Wallingford, Oxford: IOH.

Robinson, M. and K. J. Beven 1983. 'The effect of mole drainage on the hydrological response of a swelling clay soil.' *J. Hydrol.* **64**, 205–23.

Robson, J. D. and A. J. Thomasson 1977. *Soil–water regimes*. Soil survey technical monograph no. 11. Harpenden, Hertfordshire: The Soil Survey of England and Wales.

Rodda, J. C. 1969. The significance of characteristics of basin rainfall and morphometry in a study of floods in the United Kingdom. In *Floods and their computation: Proceedings of the Leningrad Symposium 1967*. IASH/UNESCO/WMO Publication, 834–43. Paris: UNESCO.

Rodda, J. C., R. A. Downing and F. M. Law 1976. *Systematic hydrology*. London: Newnes-Butterworth.

Rossi, P. H., J. D. Wright and E. Weber-Burdin 1982. *Natural hazards and public choice: the state and local politics of hazard mitigation*. New York: Academic Press.

Rostomov, C. D. 1969. Method of estimating storm runoff from small drainage basins. In *Floods and their computation: Proceedings of the Leningrad Symposium 1967*. IASH/UNESCO/WMO Publication, 462–72. Paris, UNESCO.

Royal Commission on Land Drainage in England and Wales 1927. *Report*. Cmd. 2993. London: HMSO.

Royal Society for the Protection of Birds 1983. *Land drainage in England and Wales: An interim report*. Sandy, Bedfordshire: RSPB.

Samuels, P. G. 1981. *A computational hydraulic model of the River Severn*. Report Ex 945, Hydraulics Research Station. Wallingford, Oxford: HRS.

Samuels, P. G. and R. K. Price 1976. *A mathematical model for river flows: II. Programmer's Manual*. Hydraulics Research Station Report No. INT 128. Wallingford, Oxford: HRS.

Samuels, P. G. and R. K. Price 1981. A digital model of river valley topography. *Proc. 2nd Int. Conf. on Engineering Software*. CMC Publications.

Sandbach, F. 1980. *Environment, ideology and policy*. Oxford: Basil Blackwell.

Scott, D. A. 1980. *A preliminary inventory of wetlands of international importance for waterfowl in west Europe and north-west Africa*. IWRB Special Publn. No. 2.

Self, P. 1970. Nonsense on stilts: the futility of Roskill. *Polit. Q.* **41**, 249–60.

Severn Trent Water Authority 1977. *A unified approach to land drainage*. Birmingham: STWA.

Severn Trent Water Authority 1980. *Land drainage survey: section 24(5) Water Act 1973* (8 volumes and atlases). Birmingham: STWA.

Severn Trent Water Authority 1983a. *Soar Valley improvement scheme: economic evaluation June 1983* (with Annex). Birmingham: STWA.

Severn Trent Water Authority 1983b. *Personal communication*, 18th August 1983, with Agenda item 14 for the Regional Land Drainage Committee, 22 October 1982.

Sewell, W. R. D. 1973. Broadening the approach to evaluation in resources management decision-making. *J. Environ. Man.* **1**, 33–60.

Shaw, T. L. 1979. *An appraisal of the management of isolated undrained areas within drained areas*. A report to the Ministry of Agriculture, Fisheries and Food. Rock Cottage, Blagdon, Bristol: T. L. Shaw.

Sheail, J. and T. C. E. Wells 1983. The fenlands of Huntingdonshire, England: a case study in catastrophic change. In *Mires: swamp, bog, fen and moor*, A. J. P. Gore (ed.), 375–93. Amsterdam: Elsevier.

Shoard, M. 1980. *The theft of the countryside*. London: Temple Smith.

Simmie, J. M. 1974. *Citizens in conflict: the sociology of town planning*. London: Hutchinson.

Simon, H. A. 1957. *Models of man: social and rational*. New York: John Wiley.

Slovic, P., H. Kunreuther and G. F. White 1974. Decision processes, rationality, and adjustment to natural hazards. In *Natural hazards local, national, global*, G. F. White (ed.), 181–205. New York: Oxford University Press.

Smith, D. I. 1981. Actual and potential flood damage: a case study for urban Lismore, NSW, Australia. *Appl. Geog.* **1**, 31–9.

Smith, D. I., P. Den Exter, M. A. Dowling, P. A. Jelliffe, R. G. Munro and W. C. Martin 1979. *Flood damage in the Richmond river valley, New South Wales*. Canberra: Australian National University, Centre for Resource and Environmental Studies.

Smith, D. I., J. W. Handmer and W. C. Martin 1980. *The effects of floods on health: hospital admissions for Lismore*. Canberra: Australian National University, Centre for Resource and Environmental Studies.

Smith, I. R. 1965. *An outline of British regional hydrology*. Nature Conservancy Speyside Research Station Report. Aviemore, Invernessshire: NCC.

Smith, K. 1972. *Water in Britain. A study in applied hydrology and resource geography*. London: Macmillan.

Smith, K. and G. A. Tobin 1979. *Human adjustment to the flood hazard*. London: Longman.

Snyder, F. F. 1938. Synthetic unit – graphs. *Trans. Am. Geophys. Union* **19**, 447–54.

Sokolov, A. A. 1969. The essence of the problem and the significance of the Symposium. In *Floods and their computation, Proceedings of the Leningrad Symposium 1967*. IASH/UNESCO/WMO Publication, 671–80. Paris: UNESCO.

Somerset County Council 1983. *Somerset Levels and moors plan*. Taunton: SCC.

Sorkin, A. 1982. *Economic aspects of natural hazards*. Lexington, Massachusetts: Lexington Books.

Southern Water Authority 1979. *Water Act 1973, section 24/5 land drainage survey*, 4 volumes and map sets. Worthing: SWA.

Steers, J. A. 1953. The east coast floods, January 31–February 1, 1953. *Geograph. J.* **119**, 280–98.

Sterland, F. K. 1973. An evaluation of personal annoyance caused by flooding. In *Proceedings of a symposium on economic aspects of floods*, 21–32. London: Middlesex Polytechnic.

Stroebe, W. and B. S. Frey 1982. Self-interest and collective action: the economics and psychology of public goods. *Br. J. Psychol.* **21**, 121–37.

Sutcliffe, J. V. 1978. *Methods of flood estimation: a guide to the flood studies report*. Report 49. Wallingford, Oxford: IOH.
Sugden, R. and A. Williams 1978. *The principles of practical cost–benefit analysis*. Oxford: Oxford University Press.
Suthons, C. 1963. Frequency of occurrence of abnormally high sea levels on the east and south coasts of England. *Proc. Inst. Civil Engrs* **25**, 433–49.
Swales, S. 1982. Environmental effects of river channel works used in land drainage improvement. *J. Environ. Man.* **14**, 103–26.

Thames Water Authority 1978. *Report of survey – 1973 land drainage*. Reading: TWA.
Thomas, G. J., D. A. Allen and M. P. B. Grose 1981. The demography and flora of the Ouse Washes, England. *Biol. Conserv.* **21**, 197–229.
Thompson, P. M., C. H. Green, D. J. Parker and E. C. Penning-Rowsell 1983. *A glossary of terms in flood alleviation cost–benefit appraisal*. London: Middlesex Polytechnic Flood Hazard Research Centre.
Torry, W. I. 1979. Anthropological studies in hazardous environments: past trends and new horizons. *Curr. Anthrop.* **20**, 517–40.
Townsend, J. 1980. *Storm surges and their precasting*. Storm Tide Warning Service, Bracknell: Meteorological Office.
Trafford, D. B. 1972. *Agricultural land drainage* (mimeograph). Harpenden, Hertfordshire: Field Drainage Experimental Unit.
Trafford, D. B. 1977. Recent progress in field drainage: I. *J. R. Agr. Soc. Engl.* **138**, 27–42.

Undrell, S. 1980. *The North Thames defences in the Thames Water Area*. Paper presented to Institution of Water Engineers and Scientists, South East Section, 23 April 1980.
United States Army Corps of Engineers 1978. *Rowlett Creek expanded floodplain information study, Dallas and Collin Counties*. Fort Worth, Texas: USACOE.
United States Department of Agriculture 1957. *Hydrology: engineering handbook, section 4, supplement A*. Washington: USDA Soil Conservation Service.
United States Water Resources Council 1979. *A unified national program for floodplain management*. Washington: USWRC.

Van Oosterom, H. 1981. General use and abuse of the Flood Studies Report in the United Kingdom. In *Flood studies report – five years on*, 1–6. London: ICE.
Victorian Water Resources Council 1978. *Floodplain management in Victoria*. Melbourne: VWRC.

Wakelin, M. J. 1980. *The water industry view of land drainage*. Paper presented at the seminar on 'Land drainage – whose responsibility?' London: Institution of Municipal Engineers.
Ward, R. C. 1975. *Principles of hydrology*, 2nd ed. London: McGraw-Hill.
Ward, R. C. 1978. *Floods: a geographical perspective*. London: Macmillan.
Water Act 1973. London: HMSO.
Water Resources Act 1963. London: HMSO.
Water Space Amenity Commission 1978. *Conservation and land drainage guidelines: draft for consultation*. London: WSAC.
Water Space Amenity Commission 1980a. *Conservation and land drainage guidelines*. London: WSAC.
Water Space Amenity Commission 1980b. *Conservation and land drainage working party report*. London: WSAC.

Water Space Amenity Commission 1983. *Conservation and land drainage guidelines*, 2nd ed. London: WSAC.

Welsh Water Authority 1977. *Worked examples of flood calculations, Volume 1: Rural catchments*. Glamorgan River Division. Brecon: WWA.

Welsh Water Authority 1979. *Water Act 1973, Section 24/5 land drainage survey*. Brecon: WWA.

Wessex Water Authority 1979. *Land drainage survey report, Water Act section 24/5*, 3 volumes and atlases. Bristol: WWA.

Westmacott, R. and T. Worthington 1974. *New agricultural landscapes*. Cheltenham: Countryside Commission.

Weyman, D. R. 1975. *Runoff processes and streamflow modelling*. Oxford: Oxford University Press.

White, G. F. 1945. *Human adjustment in floods*. Research paper no. 29. Chicago: Department of Geography, University of Chicago.

White, G. F. (ed.) 1961. *Papers on flood problems*. Research paper no. 70. Chicago: Department of Geography, University of Chicago.

White, G. F and J. E. Haas 1975. *Assessment of research on natural hazards*. Cambridge, Massachusetts: MIT Press.

Wilkins, J. L. 1980. Land drainage legislation and the engineer – a review and discussive paper, Parts I and II. *Chartered Municipal Engineer* **107**, 123–9, 147–54.

Wilkins, J. L. and C. J. Lucas 1980. *Land drainage legislation*. Paper presented at the seminar on Land Drainage – Whose Responsibility? London: Institution of Municipal Engineers.

Williams, H. B. 1964. Human factors in warning-and-response systems. In *The threat of impending disaster: contributions to the psychology of stress*, G. Grosser, H. Wechster and M. Greenblatt (eds). Cambridge, Massachusetts: MIT Press.

Wisdom, A. S. 1975. *The law of rivers and watercourses* 4th ed. Croydon: D. R. Publications.

Wolf, P. O. 1966. Comparison of methods of flood estimation. In *River flood hydrology*, 1–23. London: Institution of Civil Engineers.

Wood, M. 1981. A future for others? *Natural World* **2**, 14–16.

Wood, R. J. C. 1980. *The local authorities and land drainage – a district viewpoint*. Paper presented at the seminar on Land Drainage – Whose Responsibility? London: Institution of Municipal Engineers.

Young, C. P. and J. Prudhoe 1973. *The estimation of flood flow from natural catchments*. Transport and road research laboratory report No. LR 565. Crowthorne: TRRL.

Zimmerman, E. W. 1951. *World resources and industries* (Revised edition). New York: Harper and Brothers.

Author index

Numbers in italics denote figures.

General index

Numbers in italics denote figures.